D0207865

Electrical Safety in the Workplace

Electrical Safety in the Workplace

Ray A. Jones, P.E.
Jane G. Jones

National Fire Protection Association
Quincy, Massachusetts

Product Manager: Brad Gray
Production Services: Colophon
Interior Design: LeGwin Associates
Composition: LeGwin Associates
Cover Design: Cameron, Inc.
Manufacturing Buyer: Ellen Glisker
Printer: R. R. Donnelley

Copyright © 2000
National Fire Protection Association, Inc.
One Batterymarch Park
Quincy, Massachusetts 02269

Notice Concerning Liability: Publication of this work is for the purpose of circulating information and opinion among those concerned for fire and electrical safety and related subjects. While every effort has been made to achieve a work of high quality, neither the NFPA nor the authors and contributors to this work guarantee the accuracy or completeness of or assume any liability in connection with the information and opinions contained in this work. The NFPA and the authors and contributors shall in no event be liable for any personal injury, property, or other damages of any nature whatsoever, whether special, indirect, consequential, or compensatory, directly or indirectly resulting from the publication, use of or reliance upon this work.

This work is published with the understanding that the NFPA and the authors and contributors to this work are supplying information and opinion but are not attempting to render engineering or other professional services. If such services are required, the assistance of an appropriate professional should be sought.

NFPA No.: IESW-00
ISBN: 0-87765-452-2
Library of Congress Card Catalog No.: 00-105895

Printed in the United States of America
10   9   8   7   6   5   4   3   2   1

Contents

Preface

The concept of energy from electricity is relatively new in historical perspective. Michael Faraday experimented with electromagnetic induction in the early 1830s. In 1879, Thomas Edison worked out a successful principle for the electric incandescent light, using a filament of carbonized thread. In the late 1800s, George Westinghouse introduced alternating current for electric power transmission.[1] Then came the electric motor, opening the door for a new era of technology. Since that time, much work has been done on *equipment* that uses electricity. In the past century, tremendous technological strides have been made.

However, this explosion of invention in technology has come at the expense of many lives and limbs as more uses for electricity have been found. The interest in technology—invention of machines to solve problems and meet specific needs—has far exceeded the interest in safeguarding the lives of people exposed to the hazards of electric power. Historically, employers often permit their employees to be in hazardous electrical situations with little regard for the personal safety of the workers.

Electrical safety concerns developed from a natural caution around electricity because of the apparent danger. Little was known about electricity itself; thus, few plans for safe practices were made for its use. The following excerpt was taken from a booklet written in 1888:

> You all know that we have the telegraph, the telephone, the electric light, electric motors on street cars, electric bells, etc., besides many other conveniences which the use of electricity gives us.
>
> Everyone knows that, by the laws of multiplication, twice two makes four, and that twice two can never make anything but four. Well, these useful inventions have been made by applying the *laws of electricity* in certain ways, just as well known, so as to enable us to send in a few moments a message to our absent friends at any distance, to speak with them at a great distance, to light our houses and streets with electric light, and to do many other useful things with quickness and ease.

> But you must remember that we do not know what electricity itself really is. We only know how to produce it by certain methods, and we also know what we can do with it when we have obtained it.[2]

The one overriding fact that was, and is, known about electricity is that exposure to its hazards can kill. That one fact must cause employers to be concerned for the safety of their employees and cause workers to be concerned for their own safety and that of their coworkers.

Many individuals, organizations, and companies were concerned about electrical safety in earlier years. The National Fire Protection Association (NFPA) has acted as sponsor of the *National Electrical Code®* (*NEC®*)[3] since 1911. The *NEC* has been approved by the American National Standards Institute (ANSI) since 1920. The original *NEC* document was developed in 1897 as a result of the united efforts of various insurance, electrical, architectural, and allied interests. It was driven by economic loss from large numbers of fires in electrical applications installed by unskilled workers. For example, some applications used wire that was too small, and few workers knew how to make wire joints and splices. The *NEC* is purely advisory as far as NFPA and ANSI are concerned, but it is offered for use in law and for regulatory purposes in the interest of life and property protection.

In this day and age, most employers and business owners strive to make electrical safety an integral part of their operations. If all owners and employers were concerned *enough*, there would be no need for governmental intervention to protect workers. In the United States in the early and mid-1900s, only a few companies developed and relied on internal safety policies, standards, or regulations; therefore, the Occupational Safety and Health Association was born of government.

On April 28, 1971, the President of the United States, Richard Nixon, signed the Williams-Steiger Occupational Safety and Health Act. The basic purpose of The Act was to improve working environments in the sense that they impair, or could impair, the lives and health of employees.

The Act did the following:

- Authorized the U.S. government to set and enforce safety and health standards for all places of employment affecting interstate commerce

- Included protection against the hazards of electrical injury or electrocution
- Authorized enforcement officers to inspect workplaces
- Provided funding for research and education to foster work protection studies

The Williams-Steiger Occupational Safety and Health Act became known as the OSH Act, frequently referred to as The Act. With this Act, the mandate of the Congress of the United States was to assure every working man and woman in the nation safe and healthful working conditions.

As defined in The Act, the U.S. Occupational Safety and Health Administration (OSHA) initially embraced the *NEC* as the standard covering pertinent information.[4] However, the *NEC* combines fire protection and safety information.

The First Version of NFPA 70E

In the mid-1970s, OSHA determined that the *NEC* did not adequately address electrical safety as it related to *people* (the *NEC* is used only to cover installation), so the administration asked the NFPA for help. As a result, in January 1976, the NFPA assembled and chartered a technical committee to review the content of the 1972 *NEC* and to write a national consensus standard covering only the *personal safety aspects* of interacting with electrical energy.

The document that the committee produced was called NFPA 70E, *Standard for Electrical Safety Requirements for Employee Workplaces.* It was the first national consensus standard covering work practices.

The Revision Process

Because OSHA 29 *CFR* 1910, Subpart S, was based on the *NEC*, which has a revision cycle of three years, the *NEC* quickly became the more current document. The Act defined the revision process for promulgating OSHA rules as a very open and public process. In order to ensure that all elements of the public have both the right and the opportunity to participate in the process, the OSHA revision process (known as the 6(b) process for its location in The Act) potentially requires several years.

OSHA determined that the technical complexity of the *NEC* was difficult for both the employer and the employee to understand. It also determined that requirements related to work practices during operation and maintenance of electrical systems are not found in the *NEC*.

To maintain consistency with consensus standards, OSHA personnel returned to NFPA management and requested that the NFPA 70E committee be reconvened and chartered to update, revise, rewrite, and develop new sections of NFPA 70E. In addition to adding content for new issues or concerns, OSHA hoped that the updated and revised text would align more closely with the electrical safety-related work practices in OSHA Subpart S. Specifically, OSHA asked the NFPA to add information covering the following:

- Electrical protective clothing
- Electrical safe approach distances for all electricians
- Safe work practices
- Training requirements

In addition to those specific revisions identified by OSHA, the NFPA 70E committee recognized the need to do the following:

- Shift the focus of the NFPA 70E content toward people
- Improve the content of Part III, covering maintenance practices
- Develop Part IV, covering special industries

The product of the NFPA 70E committee work was the NFPA 70E, 1995 edition. The Board of Standards Review of the American National Standards Institute adopted the product as the American National Standard covering electrical safety-related work practices in 1996.

As users of NFPA 70E gained significant experience with both the language and the content of the document, technical committee members recognized a need to review the content against the experience. In addition, continuing research had expanded the electrical safety knowledge base. The fact that knowledge and experience increase constantly and the working environment changes regularly suggested that 70E should be placed in a regular review cycle.

The 2000 edition of NFPA 70E contains four distinct parts. Part I is aligned with the 1999 edition of the *NEC*. Part II is based on knowledge and experience of the various segments represented by technical committee members. Part II also embraces several new methods to estimate the degree of arc-flash hazard. Part III is revised to correlate with NFPA 70B, *Recommended Practice for Electrical Equipment Maintenance*, 1998 edition, which is the national consensus standard for electrical equipment maintenance. Part IV includes electrical safety-related work practices for special industries, such as telecommunications and electrochemical processes.

The NFPA 70E technical committee reviewed the text of the 2000 edition for alignment with OSHA 1910 Subpart S. The revision contains several adjustments in the text to achieve better alignment. However, some differences remain, due to a charter difference between the two organizations (OSHA and NFPA). The Act limits the basis for OSHA regulations to employers. Because NFPA products are developed through consensus, their products do not have a similar limitation. Consequently, NFPA 70E defines expectations of employers and employees as well as "fee" employees (contractors).

Notes

1. Parker, Steve. *Thomas Edison and Electricity.*
2. Meadowcroft, W. H. *The ABC of Electricity.*
3. The *NEC*® is a product of the National Fire Protection Association.
4. The OSH Act: Public Law 91-596.

Electrical Safety
in the Workplace

■ Chapter 1

Much Ado about a Silent Hazard

Electrical energy is one of the most powerful forces on earth. The more that is learned about it, the more impressive are its possibilities, both positive and destructive. Why, exactly, should emphasis be placed on improving electrical safety? Here are the facts:

- An estimated average of 155,000 residential electrical fires claim more than 700 lives, cause 6,800 injuries, and result in more than $1 billion in personal property damage each year.[1]
- An average of more than 4,000 nondisabling and more than 3,600 disabling electrical contact work-related injuries are recorded annually in the United States.[2]
- One person is electrocuted in the home every 36 hours,[3] and one person is electrocuted at the workplace every day.[4]
- Electrocutions were the fourth leading cause of traumatic occupational fatalities from 1980 through 1989.[5]
- The 5,348 deaths caused by electrocutions accounted for 7 percent of all fatalities and an average of 411 deaths per year from 1980 through 1992.[6]

These statistics clearly show the dangers of electricity and the importance of planning for its safe use. Although planning for electrical safety in the home is extremely important, this book deals primarily with electrical safety in the workplace. The most important factor this book emphasizes is that electrical safety in industry is a *people* issue. Planning for the *person* to be able to work safely is the key.

In most instances, electrical industry standards are related to equipment and installations. Employers and apprentice programs tend to teach and implement procedures and practices that rely strongly on the equipment and physical circumstances. Equipment construction and installation integrity are important, because some injuries are related to equipment or installation.

However, equipment construction and installation are effective only when conditions are normal. The vast majority of injuries occur when conditions are *not* normal. In normal conditions, electrical equipment tends to require little involvement from people, and exposure to a potential electrical hazard is limited. When something goes wrong, however, exposure to potential hazards increases.

What Is a Safe Workplace?

A safe workplace is one where exposure to hazards is avoided or minimized. Without exposure to a hazard, no injury can occur. In a safe workplace, an atmosphere of safety is created through a constant awareness of safety and by educating people to know how to avoid or minimize personal exposure. Elements of a safe workplace include the following:

- An electrical safety program
- Responsibility and accountability
- Recognition of factors of influence
- Measurement systems

An Electrical Safety Program

To achieve a safe workplace, an employer must institute an electrical safety program. The program should be one part of a more general safety program. Injury from an electrical hazard should be of no greater concern than injury from any other source of energy. The program may vary in significant ways, but an effective program must be considered, defined, written and published, audited, and enforced. In addition, the program should change as conditions in the workplace change, and it should be planned to provide for training and awareness.

Electrical hazards normally are not visible. Shock that results from contact with a source of electrical energy is likely to be either a "noninjury" (from a historical viewpoint) or a fatality. Although most electrical shocks do not result in injury, some do. Electrical shocks are the fourth leading cause of industrial fatality, and many other injuries occur as a result of other hazards associated with electrical energy.

Eliminating exposure to electrical hazards requires focused and continuous attention. An overall electrical safety program should emphasize specific areas of concern. The program should be well thought out and based on required tasks. People who are well versed in the subject being considered should write the program. Program authors should include safety professionals, technical professionals, and practitioners. The program should be published and readily available to all employees.

Training is a key element in an effective electrical safety program. From both a legal and an effective point of view, records of the training are important. Training should be based on the program and procedures in place within an organization. The training should focus first on building knowledge and understanding of electrical hazards, then on how to avoid exposure to those hazards. As a person completes a specific segment of training, a record should be established and maintained.

Responsibility and Accountability

The most persuasive element of any effective safety program is the responsibility felt and displayed to employees by the highest-level management personnel. A high-level manager either encourages or discourages his or her employees. Employees tend to see and hear actions much more than words. The manager is responsible for the well-being of employees whether he or she accepts that responsibility or not. The issue then becomes one of *accountability*.

In most instances, accountability is accepted as being assigned from supervision. In fact, in performance reviews and salary administration reviews, accountability normally is recognized as perhaps the most important element of a person's performance. Unfortunately, with the exception of the safety office, the accountability generally does not cover safety of people. Safety accountability that is "felt" extends in all organizational directions. The accountability begins with immediate reports and supervision. It then extends to colleagues, coworkers, families, and the entire community. This kind of accountability cannot be assigned. It must be accepted and embraced.

In a corporate environment, if a manager feels accountable to employees and their families, he or she places an emphasis on personal safety. Questions and discussions always include some visible element of this accountability. For instance, questions about productivity should be accompanied with questions about hazards and other safety issues.

In organizations with multiple management levels, each level must *feel accountable* for the safety of all persons who report to them. Responsibility then can be assigned. Supervisors should assign responsibility for personal safety to each member of the line organization. Each member of a line organization should be accountable for implementation of the employer's program, and each person's performance should be judged based on his or her implementation of the employer's program.

Recognition of Factors of Influence

The sophistication of the safety program usually depends on both the size of the entire organization and the safety organization. Achieving uniform adherence to policies and procedures in a large organization is much more difficult, because the number of people involved is much greater. Requirements for controls in small organizations usually are fairly simple. For example, the training requirements needed to implement a work practice or procedure are small, but as the size of the organization increases, training becomes increasingly difficult. Needs and requirements are independent of the organization size. However, training and communication become more complex and difficult for larger organizations.

An electrical safety program exists only if it is written and available to all employees of the corporation. The program sends to all employees a message—one that is either encouraging or discouraging. On this point, there is no middle ground. If the safety program is generated with input from all elements of the organization, the program is likely to address issues and concerns held by all levels of the organization, and the result will be positive. If, on the other hand, the safety program covers only legal and management issues, the communicated message is that the employer does not care about the individual employee, regardless of whether or not it is true. Employee involvement in the definition of the program is important.

The program content should be reviewed on a reasonable interval. The review should consider the following questions:

- Are all the elements in the program still needed?
- Are any new elements needed?
- Have working conditions changed?
- Has experience changed over time?

One element of the program should be to determine not only that there will be a program review, but how often it will take place. As conditions and equipment change, the flexibility built into the program should keep the program current. Of course, should the program change for any reason, all employees must be notified about the change.

Measurement Systems

In the U.S. system of voluntary standards, dollars normally are a significant force that drive both the existence and changes of the standards. Production of common standards eliminates a competitive disadvantage for most companies. Almost all consensus organizations come into being to save money. For example, insurance companies need a way to ensure that costs and payouts are connected to each other. Therefore, companies having the most injury claims should pay the highest rates, whereas companies having the fewest claims should pay the lowest rates.

Of course, a large employer may have many different sites and many different organizations in place on various sites. Insurers are interested only in the injury rate for an employer. However, the employer should be interested in the injury rate on each site. Although measuring injuries is one way to identify a problem, it is also seen as a way of measuring program failures. This information is satisfactory for the insurer. However, for an employer interested in the well-being of his or her employees, measuring the number of employees injured is too late, but that measurement is better than no measure at all.

An employer should design and implement a measurement system that indicates the safety program's effectiveness. The employer should consider the element for measure. The selected element might be unique to the employer, or it might be one of the generally accepted elements. For example, if an employer chooses "major injury" for focus, the term must be defined. It is likely that the insurance company defines the element to be measured and also identifies the frequency and method of collecting the information.

One example of a selected measurable element is the total recordable injury rate (TRIR). Insurance companies often establish their charges based upon accepted, measured elements. In this instance, the measured element is "recordable injury." In 29 *CFR* 1904.12, Occupational Safety and Health Administration (OSHA) enables a degree of standardization by defining recordable occupational injuries or illnesses as follows:

(c) "Recordable occupational injuries or illnesses" are any occupational injuries or illnesses which result in:
  (1) Fatalities, regardless of the time between the injury and death, or the length of the illness; or
  (2) Lost workday cases, other than fatalities, that result in lost workdays; or
  (3) Nonfatal cases without lost workdays which result in transfer to another job or termination of employment, or require medical treatment (other than first aid) or involve: loss of consciousness or restriction of work or motion. This category also includes any diagnosed occupational illnesses which are reported to the employer but are not classified as fatalities or lost workday cases.

By accepting the OSHA definition of a recordable injury, defining the measured element is easier. Insurance companies accept the OSHA definition of an injury because incidents involving these injuries normally cost money, which is important to an insurer. The TRIR is defined as the number of injuries an employer has over the course of a person-year of exposure. The number is calculated by taking the total number of recordable injuries and illnesses and dividing that number by 200,000 hours, which is about one year of exposures, as follows:

$$TRIR = \frac{\text{Total no. of recordable injuries and illnesses}}{200,000 \text{ hours}}$$

Using this equation, members of the line organization can simply review the current TRIR for one measurable indication of their experience. Because the TRIR is "post incident," it is only an indicator of the employer's experience. Measuring the health of the electrical safety program, which is covered in Chapter 19, will require a different measure—one that predicts likely injury experience.

To act on his or her accountability, a manager or supervisor participates in the measurement process. He or she is continuously aware of what the TRIR is and the experiential direction of the number. Because a higher number means a greater number of injuries, the manager or supervisor asks questions and reviews work practices and working conditions in an attempt to correct the message indicated by the number.

Why Is a Safe Workplace Needed?

The goal and, indeed, the positive results of maintaining a safe workplace are readily apparent: no injuries to people and no loss of equipment or production. More detailed objectives may be found in the following factors:

- Legal aspects
- Economics
- The right thing to do

Legal Aspects

In The Act, OSHA is chartered to establish requirements for employers. OSHA has no jurisdiction to assign responsibilities to employees. Therefore, meeting requirements defined by OSHA is the responsibility of the employer. In one view, the employer is the person who employs people. It is the employer, then, who must provide a safe workplace. It is the employer who must establish and implement a safety program. It is the employer who must establish an enforcement policy to ensure that employees follow established practices.

In the case in which a contractor is performing work on a site or facility owned by someone else, OSHA also assigns responsibilities to the facility owner. The "owner" is responsible for ensuring that the contractor (employer) sufficiently understands hazards present on the site or facility.

National consensus standards are not constrained. As a result, NFPA 70E, *Standard for Electrical Safety Requirements for Employee Workplace*, 2000 edition, also assigns responsibility. Responsibility assigned to the employer in NFPA 70E is the same as that in Subpart S of 29 *CFR* 1910. That responsibility is to define and implement an electrical safety program. The employer must generate procedures for implementation by employees. The employer must also train employees to understand and implement those procedures. Employees must then implement the procedures.

Economics

Spending money to avoid safety incidents and injuries obviously helps to avoid costs associated with accidents, such as the direct costs of repair or replacement of the failed equipment and production loss due to the failure. Indirect costs include the following:

* Incident investigation
* Creation and maintenance of a "someday" file for legal purposes
* Insurance
* Ineffective work, as employees talk about the incident
* Management reviews and reports
* Identification of procedural shortcomings and enacting "fixes"
* Legal expenses
* Medical costs

Dollars spent on an effective safety program are reported to be an excellent investment. In fact, money invested in a safety program reportedly results in a 400 percent return on investment.[7]

The Right Thing to Do

In the discussion of responsibility and accountability, it becomes clear that focusing on safety of employees is based on and, indeed, *requires* an attitude of caring. The employer is concerned for the safety of his or her employees, and the employees are concerned for the safety of their coworkers as well as themselves. If all employers and employees were concerned *enough*, no laws would be necessary.

NFPA 70E suggests that employees are responsible for implementing the procedures and the program provided by the employer. The standard goes on to suggest that although responsibilities of the employer and employee are distinct and clear, the most effective process is to establish a close working relationship between employer and employee in which each has value for the other as they work together.

This attitude of value for human life and safety provides the basis for a safe workplace simply because it is the right thing to do.

About This Book

Electrical Safety in the Workplace is designed to help owners, managers, and especially workers make electrical safety a personal priority. Education and training about risks and how to deal with hazards are extremely important. The book focuses on prevention of electrical accidents and injuries through training, understanding, and principle-based behaviors. The book can help people understand the risks and become empowered with an attitude of caring for their own safety and that of their coworkers.

Each chapter includes a section entitled "A Closer Look" that provides an accounting of an actual incident for discussion on a more personal level. In the accounts of the incidents, all names have been changed. A section of questions (and answers), called "Test Your Thinking," is also provided at the end of each chapter to aid in comprehension measurement and discussion.

 # A Closer Look

DNAS Corporation was admired for its safety statistics.[8] The company had received awards from virtually every national and international safety organization. In many countries, the federal government had recognized DNAS for its leadership in avoiding injuries. Of course, DNAS reveled in the admiration of its competitors.

The DNAS safety program resulted directly from the efforts of its founder, who was concerned about the well-being of his coworkers (employees). The safety program had helped the company to be successful. DNAS had learned that an effective safety program has economic benefits, and high value for personal safety was well established in the corporation.

But times were changing. Competition was getting more direct, development costs were skyrocketing, and profits were dropping. The company had to lower costs by reducing the number of employees. However, the company felt that safety and the environment had to remain the top priority. DNAS had successfully maintained its leadership in safety in earlier reductions in the number of employees. This time it would be no different.

However, as the number of employees was reduced, the experience level of workers, supervisors, and managers was also reduced. Demands on workers' time increased significantly.

Roger, the first-line supervisor of the power distribution crew, began the second Tuesday in August like any other day. His mind automatically thought: Check the board to see what work orders had been started on the midnight shift that would need attention during the day. He also planned to check the in-box to see what new work orders had been authorized but not yet started. Today was the first time since June that no one was on vacation.

Because today's crew was complete, Roger decided to take unit substation 3-4A off line. He would ask Eddie and Larry to transfer the load from substation 3-4A onto 3-4B and lock out the substation for maintenance. Roger would have plenty of people today to vacuum all the compartments and check bolted connections for tightness. The transformer would be heated up again before the day was over so that it would be available for use by the next shift.

Eddie knew what to do. In fact, he and Larry had written the procedure for transferring load from one substation to another. Roger talked as though there were two discrete substations. In reality, substations 3-4A and 3-4B were two separate 2,000 kVA transformers supplying separate bus for a common line of six secondary circuit-breaker units.

Each secondary circuit-breaker unit had space for one circuit breaker on each bus. Normally only one breaker would be installed in each compartment, because only one load is supplied from each compartment. Roger had made certain that each bus section was isolated from the other. Bus B was on top, and bus A was near the bottom.

A tie breaker enabled the two buses to be paralleled in order to enable on-line load transfer. Both buses were in phase and in sync.

Eddie knew what he and Larry would have to do. There were only two breakers closed on bus A. They would have to close the tie breaker, insert the spare breaker on the top (bus B), close the spare breaker, then open the bottom breaker (bus A). They would then remove the circuit breaker that was just opened and use it as a spare for the second circuit breaker. Both Eddie and Larry knew that while the transfer breaker remained closed, the amount of available energy was twice the normal amount. They had carefully selected the amount of arc-flash protection. Each person was wearing two layers of protective clothing. They also were wearing the latest in face- and head-protective equipment.

They began the process of closing, opening, and transferring circuit breakers. Each circuit-breaker handle protruded through the door of the circuit-breaker compartment. After the breaker was removed, a metal plate was to be installed to close the hole left after removing the breaker. The metal plate fit over two studs on the inside of the door. The 10-32 nuts for the metal plate were missing. Eddie rested the metal plate on the studs and gingerly closed and latched the compartment door while Larry was moving the circuit breaker to the next position. The load transfer was completed just as Roger came into the room to check on the progress. The entire process had taken about 20 minutes.

When Roger came into the room, Larry told him about the missing 10-32 nuts. Roger said that the nuts were necessary and asked Larry to go to the storeroom to get some. Roger and Eddie continued to check meters, getting ready to deenergize transformer A.

Larry returned to the substation and checked his arc-flash protection. The DNAS procedure required arc-flash protection any time a person entered the room. He proceeded to the front of the circuit-breaker unit with the replacement nuts. Roger and Eddie were standing about 18 feet away from Larry. Larry knelt in front of the circuit-breaker compartment. When he touched the door fastener, the metal plate fell from the studs onto the phase A bus. An arcing fault was initiated to ground but rapidly escalated to a three-phase fault. The tie breaker was still closed. The transformers were both supplying fault current to the point of the fault.

The arc-flash protection worn by Larry was destroyed. Both Eddie and Roger received significant burns. Larry was in the hospital for 13 weeks. He would continue to receive burn treatment for the next 6 months. Eddie and Roger were burned less severely; however, both received third-degree burns, even with the extra arc-flash protective equipment.

The investigation showed that supervisors and managers in DNAS had been comfortable in the knowledge that employees were protected from arc flash. After all, they were supplied with the latest in personal protection. The electricians, supervisors, and managers all thought arc-flash injury would be impossible if the personal protective equipment (PPE) specified in the procedures were worn.

Had the tie breaker been opened, the PPE would have been sufficient. Had the metal plate not been hanging there, the incident would not have occurred. Had the procedure required that duration of the paralleled transformers be held to minimum time, the incident would not have happened. Had the metal plate been secured to the outside of the door, the incident would not have happened. Had the other two people in the room been farther away, only one would have been injured. Had a hazard analysis been conducted, the degree of exposure would have been apparent.

 ## Test Your Thinking

1. Which of the following is *not* an element of a safe workplace?
   a. An electrical safety program
   b. Accountability that is accepted and embraced
   c. The Total Recordable Injury Rate posted for all to see
   d. Measurement systems that indicate the safety program's effectiveness
2. Which of the following statements does *not* describe an effective electrical safety program?
   a. The program should be considered, defined, written, and published.
   b. Each electrician must write his or her own program.
   c. The program provides for training and awareness.
   d. The program is audited, enforced, and changed as conditions change.
3. Money invested in a safety program reportedly results in which of the following?
   a. A 50 percent return on the investment
   b. A 400 percent return on the investment
   c. Money down the drain
   d. Money that is better spent on insurance
4. Economics of an electrical incident include which of the following?
   a. Direct costs such as repair or replacement of the failed equipment and production loss due to the failure
   b. Indirect costs such as incident investigation, insurance, legal expenses, and medical costs
   c. Costs in time such as ineffective work as employees talk about the accident and management reviews and reports
   d. All of the above

Notes

1. The National Electrical Safety Foundation.
2. U.S. Department of Labor.
3. U.S. Consumer Product Safety Commission (CPSC).
4. Occupational Safety and Health Administration (OSHA).
5. U.S. Department of Health and Human Services, Public Health Service, Centers for Disease Control and Prevention, National Institute for Occupational Safety and Health.
6. National Institute for Occupational Safety and Health (NIOSH), National Traumatic Occupational Fatalities (NTOF) surveillance system (based on death certificates of workers 16 years or older who died from a traumatic injury in the workplace). "Workers Death by Electrocution: A Summary of Surveillance Findings and Investigative Case Reports."
7. R. L. Doughty, R. A. Epperly, and R. A. Jones. "Maintaining Safe Work Practices in a Competitive Environment," *IEEE Transactions*, 1991.
8. This account is based on an actual incident. The names, including the name of the facility, have all been changed to protect those involved. Any similarity to actual names or facilities is strictly coincidental.

Answers: 1. (c), 2. (b), 3. (b), 4. (d)

■ Chapter 2

Electrical Hazards

Hazards associated with exposure to electrical current are not fully understood. In fact, all electrical hazards may not have even been identified. Fire and electrical shock are the two hazards most people recognize. However, several other electrical hazards exist but often go unrecognized. Most electricians have at least some passing experience with flying parts and pieces; many have seen eyeglasses that are covered with pock marks. Many electricians have seen the results of an explosion inside a piece of electrical equipment. Yet few electricians appreciate the large amounts of energy involved in an unintentional release of electrical energy.

Types of Hazards

In addition to fire and electrical shock, other known hazards include arc flash and arc blast, intense light, and concentrated noise. Damage to body tissue from magnetic fields and plasma is another possible hazard about which little is known.

Injuries from electrical hazards include electrocution (which *always* means fatality) from electric shock, burns, reaction injuries such as falls caused by electrical contact, and injuries from flying parts and pieces in an electrical explosion. Other injuries from arc flash, such as damage to eyes and ears, have only recently been recognized.

Fire

Each year, 30,000 fires are recorded in the United States, and investigations have found that many of those were initiated from electrical sources. Although fire is a significant danger, this book does not intend to address that particular hazard. The primary objective of the *National Electrical Code®* is to avoid sources of ignition from electrical energy.

Electrical Shock

Electrical shock (and its potential for electrocution) has been recognized as a hazard associated with electricity since the late 1800s. In fact, the state of New York identified electrocution as a humane way to execute hardened criminals. The New York prison system designed and installed the electric chair and first used it in 1890.[1] Although it cannot be pinpointed when the general public began to identify electrical shock as a hazard, the advent of consensus standards provides a beginning to the process of maintaining personal safety.

How Shock Exposures Occur

People are exposed to electrical shock in almost as many ways as the energy source is used. However, the number of ways that current flow can be initiated through a body is limited. No electrical current can flow unless a completed circuit exists. Contact with an energized point must be made in addition to contact with another point at a different voltage. In other words, a difference of potential must exist between two points on the body.

Unfortunately, in most cases, contact with ground exists most of the time. The resistance (impedance) of that contact point is important. Concrete is conductive. Metal parts are conductive. Wet or moist earth is conductive. Shoes are conductive. Gloves are conductive. The question then becomes a matter of degree. To what degree are the shoes, the concrete, or the wet or moist earth conductive? The contact impedance of these potential contact points varies widely, due to a large number of variables. The message, then, is that people should expect that they are continuously in contact with earth. To avoid exposure to electrical shock, people must avoid contact with any open energized conductor.

Electrical shock occurs when a person makes contact with an open and uninsulated energized electrical conductor. The current drawn by a tiny 7.5-watt, 120-volt lamp, passed from hand to hand or hand to foot across the chest is sufficient to cause electrocution. Parts of the person's body become conductors. As the current flows through the human body, it follows the same laws of physics as it does when the conductor is copper or aluminum. Ohm's law (current = voltage/resistance), which says there is resistance to current flow, applies, and heat is generated as current flows through the body's resistance. The essential difference, however, is that current flow through human tissue causes significant changes to occur in the tissue.

Current Flow through the Human Body

Resistance or impedance to electrical current varies from person to person, but some characteristics may be generalized. At contact, resistance to current flow has several components. Contact resistance occurs between the person and the energized point. At initial contact, such as with a hand, this resistance is about 500 ohms. Another contact point exists where the body makes contact with earth ground, such as a foot, arm, or the other hand. The body has an internal resistance of approximately 100 ohms (see Figure 2–1).

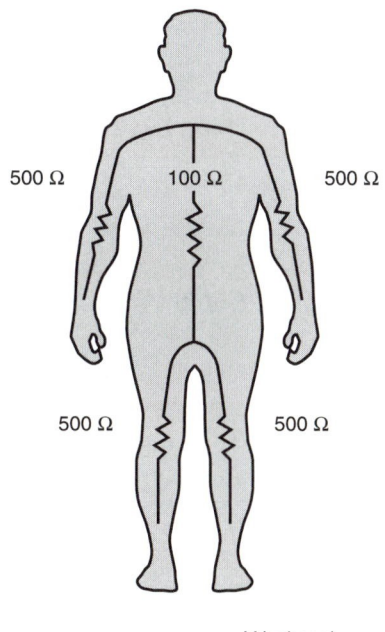

$$I \text{ (current in amps)} = \frac{V \text{ (voltage)}}{R \text{ (resistance in ohms)}}$$

Hand-to-hand = 120 V / (500 Ω + 500 Ω) = 120 mA
Hand-to-leg = 120 V / (500 Ω + 100 Ω + 500 Ω) = 110 mA

Figure 2–1. Current Flow through the Human Body.

Both blood and nerve tissue are good conductors of electricity. Blood, especially oxygenated blood, is an electrolyte, which is especially efficient in carrying electrical energy. One critical function of nerve tissue is to carry the electrical signal generated in the brain to the muscles. One prime purpose of nerve tissue, then, is to conduct small amounts of current signals to the muscles, where they can either constrict or release from constriction.

Table 2–1 illustrates how a person's body might react when exposed to electrical shock. Because the frames of women are usually smaller than those of men, women tend to suffer damage at smaller amounts of current exposure.

The table illustrates what a typical body reaction might be to current flow at various levels. The typical body reactions in Table 2–1 can be expected when the current flow lasts for about one second. Current flow is unlikely to remain constant because it increases as the duration of the contact increases. Note that all entries on the chart are listed in amperes.

The amount of electrical current on each line of Table 2–1 is extremely small. An ordinary light bulb requires about 1 ampere. When a shock of less than 0.010 amps of current is received, the person probably experiences a sensation of "being pulled into the circuit" and may feel some pain.

Table 2–1
Reaction of Human Body to Electric Current

Effect of Current	AC Current in Amps—Men	AC Current in Amps—Women
Perception threshold (tingling sensation)	0.0010	0.0007
Slight shock—not painful (no loss of muscle control)	0.0018	0.0012
Shock—painful (no loss of muscle control)	0.0090	0.0060
Shock—severe (muscle control loss, breathing difficulty—onset of "let-go" threshold	0.0230	0.0150
Possible ventricular fibrillation (3-second shock)	0.1000	0.1000
Possible ventricular fibrillation (1-second shock)	0.2000	0.2000
Heart muscle activity ceases	0.5000	0.5000
Tissue and organs burn	1.5000	1.5000

Source: U.S. Department of Energy *Electrical Safety Guidelines*, Appendix A, Sept. 1993.

As current flow increases from 0.023 to 0.040 amps, the signal received by the muscle tissue from an external source of energy begins to exceed the ability of the brain to tell the muscle tissue to release. The stronger signal is for the muscle to constrict. The "let-go" threshold has been reached. On occasion, the external signal overcomes the ability of the body to release the contact. At this point, only an external force can break the connection. Someone or some force must intervene. As the length of contact duration increases, the current flow increases, causing more significant impact.

When the current flow reaches about 0.200 amps and the duration of the contact increases beyond one second, fibrillation becomes a real possibility. Fibrillation is the very rapid irregular contraction of the heart muscle that occurs when the heart rhythm is interrupted. If not corrected, fibrillation will cause oxygen flow to body tissues to cease and the person to die. The likelihood of fibrillation increases as the duration of the contact increases. Yet a current of 0.200 ampere is very small, and one second is a very short period of time. Once fibrillation occurs, only medical intervention within the first few minutes can restore normal heart activity. References suggest between three and four minutes as the maximum length of time before medical intervention must occur.

Table 2–1 suggests that, essentially, only two results should be expected from contact with an exposed energized electrical conductor. One is that a person experiences a tingling sensation. When this happens, the person likely jumps back, looks around to see who saw the incident, and continues with the work, if no one was looking. The other possible result is that the current flow escalates and fibrillation is reached. The person is found lying on the floor not breathing—another fatality.

Current flow in the human body follows three common paths. Those paths are generally called hand to hand, foot to foot, and hand to foot. Figures 2–2, 2–3, and 2–4 illustrate these paths.

Although the diagrams seem to suggest that, in some instances, externally impressed current flow does not affect the heart muscle, nothing could be further from the truth. Any electrical current passed through blood and nerve tissues will be conducted to vital organs, such as the lungs, kidney, heart, and brain.

It is important to remember that any item of electrical equipment can be the source as indicated in the diagrams. Any electrical equipment with-

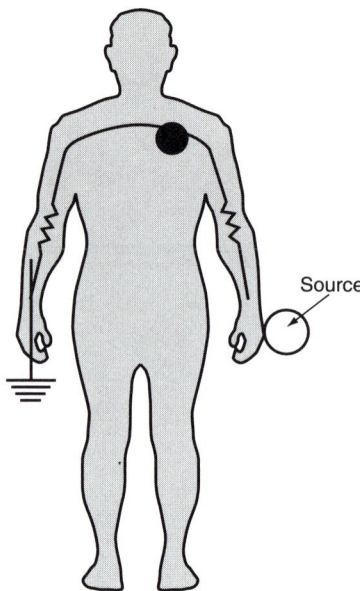

Figure 2–2. Touch Potential for Current Flow Path.

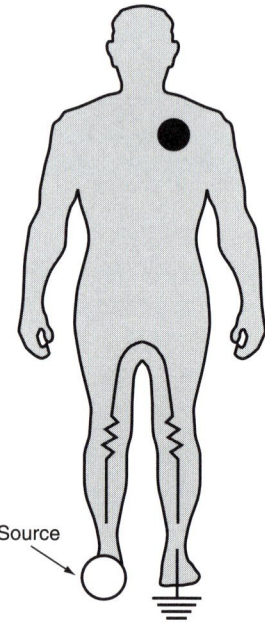

Figure 2–3. Step Potential for Current Flow Path.

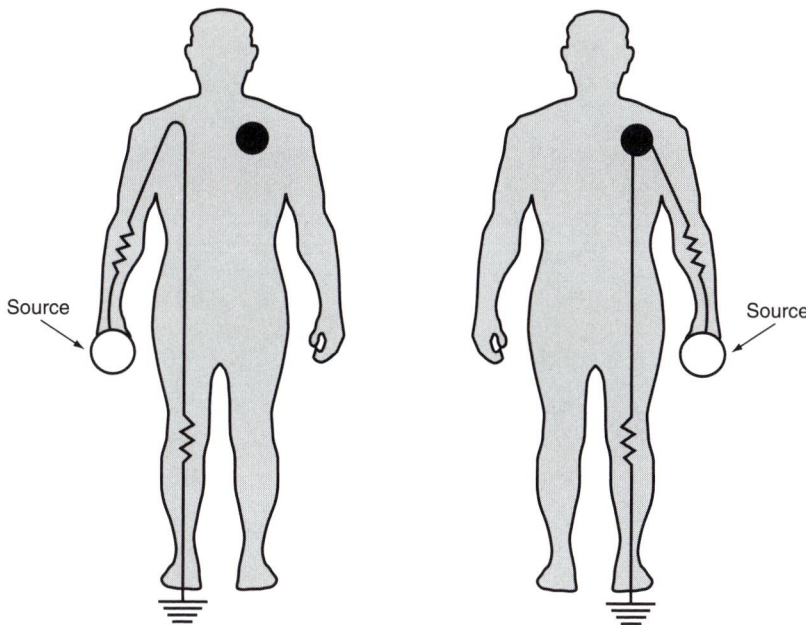

Figure 2–4. Hand-to-Foot Current Flow Path.

out adequate grounding might be the source of energy in the diagrams. The electrical equipment might be a motor, a knitting machine, an extruder, a part of the building steel, a sewing machine, or a coffeepot.

The term *live part* is used in many standards, but the term is not always used to mean the same thing. Each standard provides a definition for the term as it is used in that standard. Live part has been used for many years in the trade without consistent definition. The term connotes something living and breathing, and perhaps that connotation is not far off in the electrical sense. An energized point on an open and uninsulated conductor seems to be waiting for someone to make contact. The live part will wait for days, weeks, or months for someone to make a mistake, and when a mistake is made, the live part will strike.

To avoid electrical shock, simultaneous contact with a live part and ground must be avoided. As indicated earlier, without a completed circuit, no current can flow. However, it is possible to contact an open energized conductor without injury, as birds and squirrels demonstrate every day. On occasion, an electrical utility worker may be required to perform work on an energized service. If this work method is necessary, it should be reserved for specifically trained and qualified craftspersons in the utility distribution industry.

Workers can increase their protection by wearing equipment that serves as electrical insulation. If a person's hands and arms are fully insulated, that person possibly can touch an energized conductor directly without being injured. On occasion, electrical workers may be required to use this work method to accomplish a task, but this practice is *not* recommended. Only specifically trained, qualified people should attempt this work method, and then, only while they follow procedure controls discussed in Chapter 5, "Strategies for Preventing Injury."

Generally, exposures that are planned and executed in accordance with the plan are very successful. When a specific activity is planned with actions and reactions anticipated, injuries normally don't happen. The degree and depth of planning are extremely important, and simply reacting to each situation is an unacceptable work practice. Every action must be planned.

Arc Flash

Arc flash is a release of concentrated energy that is the result of an arcing fault. The flash results from the passage of substantial electric currents through what had been air but becomes the vapor of the arc-terminal material, usually copper or aluminum. The way that records are kept has a bearing on how hazards are understood. For instance, the medical community tends to record injuries and treatment by the type of injury. A burn will be treated as a burn, with the injury and treatment recorded in the same category as all burns. Most injury records do not identify the energy source or type of energy. One key issue with this system of record keeping is that it tends to obscure information that might be valuable to prevent the injury in the first place. Arc flash is a case in point.

Arc-flash burns have occurred since the initial widespread use of electricity as a general energy source. However, as a physician treats an arc-flash burn, the injury may be recorded as a burn. Even though an electrical arc causes the injury, people tend to think of the injury as a burn, not an arc-flash burn. Available data fail to differentiate between arc-flash injuries and other types of burns. The general community will not react until the existence of a hazard is recognized. Although a few individuals and some organizations recognized arc flash many years ago, arc flash is only now being identified as an electrical hazard.

A couple of individuals in the Institute of Electrical and Electronics Engineers (IEEE) community pioneered recognition of the arc-flash hazard in the mid-1980s. Ralph H. Lee authored a paper, "The Other Electrical Hazard," which was presented at the Industry Applications Society (IAS)/IEEE Petroleum and Chemical Industry Conference in 1981.[2] Lee's paper suggested that an arcing fault is vastly different from a bolted fault. In his paper, Lee defined a set of parameters that would cause any burn received by a person to be "curable." In other words, Lee's objective was to identify the point at which any burn would *not* be incurable. Another pioneer, Bill Jordan of Dow Chemical, established arc-flash protection schemes in his company before most people recognized the hazard.

Recognizing the Hazard

Often people and organizations will not listen to a new message, such as the one in "The Other Electrical Hazard," unless they have similar experiences. Some organizations had workers who were recently injured from arc flash. Those individuals and organizations embraced Lee's paper. The remainder of the community did not recognize the central message: "Here is something new and important." Only as more individuals and organizations experienced arc-flash burns did recognition begin to grow within the Petroleum and Chemical Industry Committee of the IEEE Industry Applications Society.

With its publication in 1995, NFPA 70E, *Standard for Electrical Safety Requirements for Employee Workplaces,* became the first consensus standard to identify arc flash as a hazard to be dealt with by the general community. Following the publication of NFPA 70E, OSHA promulgated 29 *CFR* 1910.269, which contained some recognition of the existence of a potential injury from an electrical arc. The initial acknowledgment of the existence of arc flash has generated increased interest. Both the general industry and the utility industry have shown an increased interest in knowing more about the hazard and how to protect people who may be exposed.

Examining the Equipment

Electrical equipment manufactured and installed to distribute electrical energy most likely has been tested for faulted conditions by the manufacturer. Several consensus standards address this testing process. The test is intended to mimic a faulted condition in the field and to monitor the ability of the equipment to dissipate the electrical energy without damage to the equipment. In a bolted-fault test, the primary forces being tested are mechanical forces exerted on the bus resulting from magnetic force. The test is conducted by placing a bolted fault or short circuit in the equipment. The amount of magnetic force that will be applied to bus work during a bolted fault can be calculated. In general, consensus tests are designed to test the integrity of both these calculations and the construction of the bracing holding the bus work in place.

Bolted faults do occur. On occasion, an electrical service is energized with the safety grounds in place. If the grounds are connected effectively, the resulting fault essentially duplicates the set of conditions defined by the test protocol as in ASTM F 855, 1997.[3] If the ground set is not properly constructed and installed, magnetic force could destroy the connections of the safety grounds.

No tests are performed to verify the integrity of electrical equipment during an arcing fault. There are several reasons for this omission. Equipment purchasers have not asked for it. The need to remain competitive provides disincentive for a manufacturer. The number of defining variables in an arcing fault makes it very difficult to design an effective test for an arcing fault. Arc faults are extremely unpredictable. In fact, it is only clear that the amount of energy available at the point of the fault defines the amount of destruction or injury that is precipitated in such an event.

Figure 2–5 illustrates that the point of danger from arc-flash depends on the amount of energy that is available at the point of exposure. Regardless of the method of calculating the arc-flash boundary, the point of danger varies with the amount of energy that will be released during the arcing fault condition.

Addressing the Hazard

Understanding is growing. Each year, more testing is designed and performed that adds to the bank of knowledge. The 2000 edition of NFPA 70E contains at least two ways to calculate arc-flash boundary.

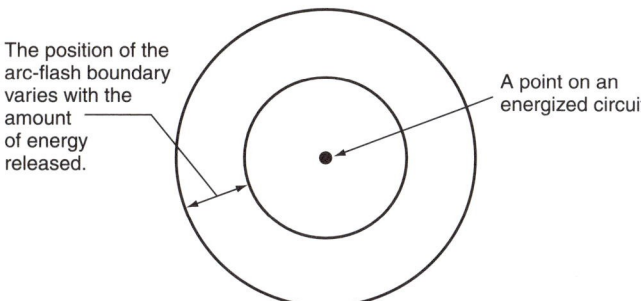

Figure 2–5. Point of Danger from an Arcing Fault–Arc-Flash Boundary.

When the technical committee on Electrical Safety Requirements for Employee Workplaces was deliberating on how to address the arc-flash hazard, opinions varied from "no hazard exists" to "the standard must provide maximum protection." The technical committee was aware that some members of the electrical community held an opinion that electricians are paid handsomely to work with electricity and should be expected to work on hazardous tasks. Even people who had personally experienced arc-flash burns had expressed that opinion. Data were provided to the technical committee that verified a significant number of burn injuries from electrical arcs. Because arc flash was not recognized formally as an electrical hazard, no consensus standard dealt with personal protective equipment (PPE). Although industry had recorded significant experience with flame-resistant clothing for other purposes, little experience existed in protecting people from arc flash.

Prior to the 1995 edition of NFPA 70E, consensus standards dealt with electrical explosions by defining equipment design, manufacture, and installation necessary to protect the equipment. An electrical explosion generally damaged equipment and lost product, both of which cost a great deal of money. Purchasers were interested in maximizing investment, so they were willing to spend some money to avoid costs associated with electrical explosion. A second issue is that consensus standards generally addressed conditions in which the equipment was operating "normally." Few requirements were associated with "abnormal conditions."

In his paper, Ralph Lee suggests that a potential arc flash exists in all instances where electrical energy is used. The primary point of the paper is that if energy in the arc flash is sufficiently high, a burn injury likely will result if a person is within the limit of the flash zone. The paper discusses variables associated with an arc flash and offers an equation to calculate the amount of energy available in the arc flash. The paper suggests that the temperature within the arc is extremely high.

The technical committee for the 1995 edition of NFPA 70E recognized that a significant number of injuries occur in industry (both general industry and the utility industry) as a result of arc flash. One major chemical company provided the technical committee with information that, with sufficient effort and PPE, arc-flash injuries could be avoided.

The committee felt compelled to include in the document information about the potential hazard and how people might protect themselves in the event of an exposure. Although lacking consensus information about PPE, a significant amount of anecdotal information was available from organizations represented by technical committee members.

Understanding Curable Burn

According to "The Other Electrical Hazard," a *curable burn* occurs if a person's skin is exposed to an accidental release of energy defined by the following equation:

$$D_c = (2.65 \times MVA_{bf} \times t)^{1/2}$$

This equation means that the *curable burn* distance is the distance, in inches, that this equation defines. The intent is that any burn experienced as a result of an exposure beyond this distance is *curable.* The term *curable burn* was generated by Lee and defined for this paper as being roughly the same as a second-degree burn as defined in the medical community.

The D_c in the equation stands for *curable burn distance.* The constant, 2.65, was generated by Lee, based upon testing that he executed.

MVA_{bf} represents the *available bolted-fault current* at the point of the failure. Bolted-fault current should be calculated with generally available methods.

Time is represented by t in the equation. Time is expressed in seconds and represents the amount of time during which the fault current flows. By combining the available bolted-fault current with time, the amount of available energy results.

After performing these number manipulations, the square root of the result is taken, and that number is expressed in inches. The result is the curable burn distance hypothesized in Lee's paper.

Members of the NFPA 70E technical committee reported some experience with arc-flash injuries and used this equation to understand the degree of exposure to arc flash. Knowledge acquired in these experiences aligned completely with the distance suggested in Lee's paper.

Understanding Incident Energy

The 2000 edition of NFPA 70E added a second idea about how to determine the degree of the arc-flash hazard.[4] A series of tests were conducted in a laboratory environment to categorize protective characteristics of clothing,[5] resulting in the generation of an empirical formula. The formula is based on the amount of *incident energy* absorbed (in human skin tissue) that results in a second-degree burn. NFPA 70E defines this amount of energy as 1.5 calories per square centimeter for the (normally) very short duration of electrical arcs.

Appendix B-5 of Part II in the 2000 edition of NFPA 70E provides the following empirical formula for arcs in open air:

$$E_{MA} = 271 \, D_A^{-1.9593} \, t_A \, (.0016 \, F^2 - 0.0076 \, F + 0.8938)$$

Where: E_{MA} = Maximum open arc incident energy, calories per square centimeter

D_A = Distance from arc electrodes, inches (for distances 18 inches or greater)

t_A = Arc duration in seconds

F = Bolted-fault short-circuit current in kA (for the range of 16 to 50 kA)

For arcs in a cubic box, the 2000 edition of NFPA 70, Part II, Appendix B-5, contains the following empirical formula:

$$E_{MB} = 1038.7 \, D_B^{-14738} \, t_A \, (0.0093 \, F^2 - 0.3453 \, F + 5.9675)$$

Where: E_{MB} = Maximum 20-inch cubic box incident energy, calories per square centimeter

D_B = Distance from arc electrodes, in inches, for distances of 18 inches or greater

t_A = Arc duration in seconds

F = Bolted-fault short-circuit current in kA (for a range of 16 to 50 kA)

These formulas provide a value for incident energy that can be used to select a protective clothing system from a table contained in the standard.

Managing the Hazard

On the basis of such results, the technical committee embraced "The Other Electrical Hazard" as the definition of when a person should be protected from the extremely high temperatures generated in an arc.

Although available fault current may be managed by controlling internal impedance or by reactors, time is the easiest variable to manage. Overcurrent protection schemes have been required in electrical circuits for many years. Existing requirements are for fuses and circuit breakers to be selected and sized to protect conductors, circuits, and equipment. The idea is that if conductors, circuits, and equipment are protected from *explosion,* then people are also protected from both fires and explosion. The concept is true, provided the equipment is operating normally and completely deenergized prior to opening the equipment enclosure. Current national consensus standards provide excellent protection where the conductors and equipment are closed and operating normally.

The problem comes when the operation changes from normal to abnormal, and a person is asked to find out what happened and to repair the problem. Usually, the first thing a repair technician does is to open the equipment to troubleshoot the problem. Such a condition is abnormal, and the risk of an injury increases. When an engineer selects and sizes overcurrent protection, he or she should account for the fact that, under abnormal conditions, exposure to arc flash increases and risk of injury increases commensurably.

Applying Overcurrent Protection

Designers and engineers should begin to think about overcurrent protection as one element of a people-protection scheme. The amount of energy released in an arc-flash event is related to the square root of the duration of the arc. There are significant implications in this concept. Faster operating equipment generally costs more. Selling the faster devices becomes a problem because equipment with increased operating speed is not required.

Arc-flash events happen extremely fast. Circuit breakers manufactured several years ago can clear a fault in 30 cycles, or one-half of one second. Currently manufactured circuit breakers typically clear a fault in 12 cycles, or one-fifth of one second. Fuses and some circuit breakers clear a fault in less than 2 cycles, or one-thirtieth of one second. One very important factor to determine the operating time of an overcurrent device depends on the *degree* of a fault. The size of a fault changes the operating time for the overcurrent device. Manufacturer's time/current curves must be consulted to accurately determine the operating characteristics of any specific overcurrent device.

Avoiding Exposure

Human reaction time varies from one person to another. However, based on a time/motion study, a typical reaction time is 0.4 second (four-tenths of one second). Reaction time expressed in cycles (at 60 Hz) is 24 cycles. This means that 0.4 second (24 cycles) is the shortest time in which a person can view a condition and begin to move or react. Therefore, in a 30-cycle fault, a person might begin to move. In all other conditions, it is not possible for a person to see an arc and move before the event is over. The only possible options are to avoid any exposure to arc flash or to wear PPE that will prevent injury. Although standards for protective clothing are being defined and issued, avoiding exposure to the hazard is the only method known to avoid injury.

Thermal injuries—burns from hot surfaces, flames, building fires— are reasonably common. The mechanics of these types of injuries are well documented and widely known. Such burn injuries are sustained as a result of exposure to sources with temperatures of several hundred degrees. Temperatures of 14,000°F are very common in electrical arcs. An arcing electrical fault can generate temperatures in excess of 30,000°F. During an arcing fault, the temperature escalates until the source of energy is removed or the maximum temperature of 30,000°F to 35,000°F is reached.

Human tissue exposed to such thermal sources is effectively burned to a crisp. If the overcurrent protection can remove the energy rapidly, the maximum temperature is limited, but temperatures of 14,000°F are still common.

Flammable or meltable clothing is likely to ignite or melt. If the clothing ignites, the duration of the exposure is long, and the injury is likely to be more widespread. Many fabrics are made from fibers that melt before igniting. In such an instance, the hot fibers land on the human tissue and literally melt into the skin surface, causing many significant injuries. Flammable or meltable clothing should not be worn when there is a potential for exposure to arc flash.

Arc-flash testing performed in the laboratory of Ontario Hydro in Toronto, Canada, attempted to quantify the protective characteristics of various pieces of protective clothing. [The American Society for Testing Materials (ASTM) Standard, ASTM F 1503, defines the testing method used in these experiments.] The results of these tests are reported in three IEEE papers.[5, 6, 7] The Ontario Hydro testing program served as the basis for the tests defined by ASTM in F 1503.[8]

The tests served as one basis to evaluate the protective characteristics of various clothing constructions. Based on these tests, the ASTM committee completed work on ASTM F 1506, 1998 edition.[9] This standard provides information on the amount of protection for different exposure levels. (See information on PPE in Chapter 5, "Strategies for Preventing Injury.")

Injuries from such very high and rapidly escalating thermal conditions tend to be medically different from other types of burns. During an arc-flash event, the arc-plasma fireball engulfing the exposed body part has a magnetic element in addition to the known thermal property. Investigations are being performed at the University of Chicago Trauma Center to gain understanding of burn injuries experienced in releases of electrical energy. The effects of the "magnetic" energy exposure are unknown at this time.

To provide people with the best chance of avoiding arc-flash injury, it is necessary to manage exposure to the hazard. One element of managing personnel exposure is to establish an electrically safe working condition (see Chapter 8, "Impacting Work Practices"). The concept of an electrical safe working condition is that all unprotected electrical energy has been removed from the work environment so that it is not practically possible for an injury to occur from electrical energy sources.

Reducing Degree of Exposure

If it is not possible to establish an electrically safe work condition, the next best approach is to reduce the *degree of exposure,* another way of saying minimize exposure to the hazard. The following methods may help reduce the degree of exposure:

- Covers and doors should stay closed and latched until the last possible task before exposing the conductors.
- The person should stand to the side of the door or cover while opening the equipment.
- The person should be wearing PPE identified in the hazard analysis (see Chapter 3, "Hazard Analysis").
- If more than one person is needed at the scene, only one should remove covers or open doors. Any other person should remain outside the flash protection boundary (see Chapter 13, "Safe Work Practices").

Experience has shown that arc-flash events are most likely when physical movement is occurring within the equipment. Arc-flash events are very rare when switches, circuit breakers, or contactors remain open or closed with equipment operating normally. That same experience shows that arc-flash events are most likely to occur in the following situations:

- When a switch or circuit breaker is closed (or opened)
- When a door or cover is opened (or closed)
- When equipment is inserted onto (or withdrawn from) the bus
- When test equipment or safety grounds are installed

Therefore, wearing PPE is most important when the degree of hazard is greater.

Arc Blast

As opposed to arc flash, which has a thermal hazard, arc blast is associated with extreme pressure and rapid pressure buildup. For example, electricians often talk about a specific piece of electrical equipment that "blew up when it was turned on." Both construction and maintenance electricians relate stories not only about the blast, but also about the amount of equipment destruction that occurred. Most engineers are aware of potential explosions. On occasion, a news program reports that an electrical explosion resulted in loss of electricity to a major building somewhere. It almost seems that the frequency of these events is increasing.

Most electricians are aware of the flying parts and pieces in an electrical explosion. However, it is difficult for most electrical professionals to grasp the amount of physical force in such an explosion.

Recognizing the Hazard

A substance requires a different amount of physical space when it changes physical state, say, from liquid to solid. Although most solid materials consume less space than they do in a liquid state, water is an exception. Water expands when it freezes. When a person fills ice trays with water and later removes the ice, he or she notices that the volume of ice is greater than that of the water. Water is an unusual material. Most solid materials consume less space than they do in a liquid state.

If the change of state is from liquid to gas or vapor, the substance often requires a great deal more space at the same pressure. For example, when water turns to steam, it expands four times.

Electrical conductors generally are made of copper. During an arcing fault, the very high temperature causes the solid copper material first to melt and then to vaporize. In the initial change of state from solid to liquid, the space consumed by the material increases slightly. When the liquid copper vaporizes, it expands 67,000 times. Although the amount of copper may be small, the extremely high expansion factor results in a powerful force.

In addition to powerful pressures generated by the change of material state, the extreme temperatures of the arc heat the air rapidly in a manner similar to a lightning stroke. Thunder is the sound created by the rapidly expanding air as it is heated by the current flow in the lightning discharge.

In most instances, arc flashes are enclosed in metal. Only in rare instances does the metal enclosure provide a path or mechanism to release the pressure generated when the copper changes state. The normal result is that a powerful force is created as the pressure builds within the enclosure. In many instances, the force increases until some component yields. Sometimes the enclosure is distorted sufficiently to relieve the pressure. Sometimes a door opens, which, in turn, exposes any person who happens to be in front of the device to arc-flash hazards. Most arc-flash events occur when something is physically happening. If the arc-flash event is the result of opening or closing a door, the person taking the action is in trouble if he or she is standing in front of the door.

Another prime initiator of arc-flash events is a person taking some action that should not be taken and most likely being positioned directly in front of the event. As indicated earlier in this chapter, an arc-flash event is now recognized as an electrical hazard. However, the intense thermal event is the only arc-flash result that is addressed in any consensus standard. Protective clothing prescribed in NFPA 70E and defined by ASTM F 1506 provides protection from the thermal aspects of the event. However, no currently viable protection scheme is in place for exposure to forces resulting from the pressure rise either in an enclosure or in the air.

A person who is exposed to the pressure wave front caused by arc blast has a differential pressure between the side of the body initially exposed and the opposite side. Although any differential pressure will equalize rapidly, internal body organs are subjected to the differential pressure. The result of this exposure is not readily understood. However, according to researchers at the University of Chicago Trauma Center,[10] significant damage to the brain or lungs is possible.

Addressing the Hazard

Some electrical enclosures are not susceptible to deformation when an arc event occurs within the enclosure. Some enclosures are made of cast iron or cast aluminum, such as a motor terminal box. In several documented instances, a motor terminal box exploded as a result of force created by the pressure buildup within the enclosure.[10]

Because the pressure wave created by an arcing fault generally is not recognized as a hazard, pressure-relief paths receive little consideration, except for explosion-proof enclosures. However, both pressure and the rate of pressure rise are much greater than most "fuel" explosions when the pressure and rate of the pressure rise are created by an arcing fault.

An exploding motor terminal box sometimes results in parts of the enclosure "flying across the room" with significant momentum. A person in the path of the "flying part" may receive a significant injury. It is possible that significant property damage also can result from this type of explosion.[11]

An event with such extreme thermal and explosive forces likely will result in intense light and sound energy. Such is the case of an arc-flash event. Little is known about this type of injury, yet people exposed to this type of event have reported vision trouble that lasts for several days. Cataracts resulting from electrical flash also have been reported up to three years after an accident.

OSHA identifies 85 dBA as the intensity that requires ear protection. (Note that the limit is related to exposure over a period of time, decreasing as the intensity increases.) Explosions from arc flash and blast can reach many times that level. Hearing damage has been documented as one result of exposure to an electrical arcing fault. Ear protection is available. However, one must choose any hearing protection with an understanding that a pressure wave front hazard also exists. Hearing protection must be worn that will not be driven into the ear by the accompanying pressure wave front.

Figure 2–6 shows actual noise and pressure wave front measurements taken in a series of arcing fault tests.

A workplace electrical accident can affect the victim's skin, ears, eyes, lungs, internal organs, and nervous, muscular, and skeletal systems. The effects are a direct result not only of electrical current but also of such things as disturbance of the heart's rhythm—barotrauma from acoustic and vibratory forces of arc blast.

An arcing fault is an extremely complex event involving different types of energy released by several means. To avoid injury, hazards must first be analyzed individually and then collectively. Little is known about how the types of energy or hazards interact with each other. When a person selects and takes protective measures, he or she must consider all hazards.

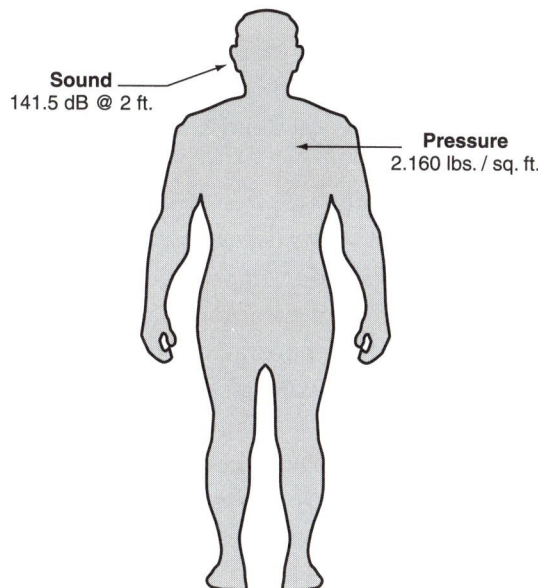

Figure 2–6. Pressure Measurements.
Source: Jones, Ray A., et al. "Staged Tests Increase Awareness of Arc-Flash Hazards in Electrical Equipment." Paper presented at the Forty-Fourth Conference of the IAS/IEEE Petroleum and Chemical Industry Committee, Banff, Alberta, Canada, September 15–17, 1997.

 A Closer Look

In an industrial manufacturing plant,[12] a continuous process is shut down for annual maintenance. One contractor plans to install and terminate some new wiring inside an existing motor control center (MCC). The wiring is being installed to enable a new monitoring system for the process. Detailed drawings have been prepared. The new cables have been pulled and left coiled up in the cable tray above the MCC.

On a Sunday morning, the electrical foreman assigns Charlie Hughes to install the cable into the starter unit and terminate the conductors. Charlie has all the required information and all the necessary tools and equipment. He is a good electrician with 17 years' experience. However, he is new to this job site, and he has never worked with this particular product line. Charlie takes the top cover off the MCC to find a good place to bring the cable into the wire way. The wire way is enclosed and has plenty of space. Charlie drills a pilot and punches a hole for a 3/4-inch conduit. He bends, cuts, threads, and installs the conduit from the wire way to the tray. He proceeds to push the cable into the wire way and thread the cable to the correct size-one starter unit.

In this particular product, the terminal block is in the vertical wire way. Charlie loosens the fastener and lifts the terminal block cover. At this point, Charlie recognizes a problem: The terminal block has no identification. The block has nine points, but none of the blocks is marked with identification numbers that are shown on the schematic diagram. How is he going to land the seven new wires? How can he correlate the conductors in the new cable with the correct terminal block?

Although all equipment normally supplied by this MCC is out of service, the MCC bus remains energized. A "spare" disconnect switch in the MCC has been selected to provide "temporary" power to the lighting transformer. It is not necessary to rent and install a generator if the MCC bus can remain energized. The site has planned for all disconnects to be opened except the one feeding the lighting transformer. Anyway, the new cables are to be landed in the wire way. The starter unit doors can all remain closed. Charlie is a "qualified person." In fact, he has received all required training and is the "hook up" person for the crew. All the documentation is in order.

Charlie is aware of the need for accuracy and completeness of this new work. When he notices the missing terminal block numbers, he calls to his foreman, Dan, to report the incomplete information. Charlie relays his observations to Dan, who says, "No real problem. We'll open the door and trace the wires. The schematic shows all the internal wiring." Dan doesn't wait for Charlie to respond. He simply opens the starter door and says, "See this point on the schematic? That's this point here in the starter." The internal wiring is neatly bundled and tied. Not wanting to cut the wire ties inside the starter, Dan begins to follow the conductors from the known component to the terminal block, with the idea that his crew can leave the terminal block identified for future reference.

As Dan traces one conductor with his thumb and forefinger of both hands, he is unaware that his left pinky has curled around behind a cover. The manufacturer had installed the cover to prevent contact with a live part. Dan's curled finger slips behind the cover and contacts an energized point. Although the disconnect is locked in the open position, the starter is still in contact with the bus. Dan touches a point electrically ahead of the disconnect switch. Due to the confined space inside the starter unit, Dan's right arm is touching the metal enclosure.

While Dan is tracing the conductors with his hand, Charlie turns to talk about the equipment with another person. When Charlie turns around, he sees that Dan is stiff and not moving. Dan makes no sound, and he does not acknowledge Charlie when Charlie calls his name. Aided by increased adrenaline, Charlie pulls Dan from the starter unit. When Charlie asks Dan if he is all right, Dan says, "I've been hit." Those are his last words. Dan has become part of the statistics that record electrocution as the fourth leading cause of industrial fatalities.

A few thoughts come to mind:

- Why not install a temporary generator and shut down the MCC bus entirely?
- Why not use a meter or some other device to trace the wires?
- Why not cut the wire ties?
- Why not stop the work when it becomes known that the identification is missing until the work is replanned?
- Why is the missing identification a surprise?
- Why not look at the terminal block the day before the shutdown?

There are no good reasons to take the chance that these workers took. A few more minutes of planning this job and changing the plan when unexpected findings arose might have saved Dan's life.

Charlie's reaction to call his supervisor seems to be appropriate. But Dan's reaction—to jump right in and trace the conductors—was not appropriate. After the accident analysis, a small burn found on Dan's left pinky indicated where the current entered his body. The current flow was so small that no identifiable point was found where the current exited.

How can exposure to electrocution be avoided?

The only real way to avoid being exposed to electrical shock is to remove the source of energy and take some action to provide assurance that the energy remains removed.

Test Your Thinking

True	False	
❑	❑	I. Known electrical hazards associated with exposure to electrical current include only fire and shock.
❑	❑	2. Electrocution means being shocked by electricity. Electrocution is sometimes, but not always, fatal.
❑	❑	3. When exposed to the same level of current flow, the bodies of men and women react differently.
❑	❑	4. Arc-flash hazard was not identified as a hazard until the mid-1980s and was only officially recognized as a hazard to be dealt with by the general community after the publication of NFPA 70E in 1995.

Notes

1. San Francisco State University Home Page (http://www.sfsu.edu/~markd/acpower.htm), April 8, 1999.
2. R. H. Lee, "The Other Electrical Hazard: Electric Arc Blast Burns."
3. ASTM F 855, *Standard Specification for Temporary Grounding Systems to be Used on De-energized Electrical Power Lines and Equipment,* 1997.
4. NFPA 70E, *Standard for Electrical Safety Requirements for Employee Workplaces,* 2000 edition. Part II.
5. R. L. Doughty, T. E. Neal, and H. L. Floyd, "Predicting Incident Energy to Better Manage the Electrical Arc Hazard on 600 V Power Distribution Systems."
6. T. E. Neal, A. H. Bingham, and R. L. Doughty, "Protective Clothing Guidelines for Electric Arc Exposure."
7. R. L. Doughty, et al. "Testing Update on Protective Clothing and Equipment for Electrical Arc Exposure."
8. ASTM F 1503, *Standard Practice for Machine/Process Potential Study Procedure.*
9. ASTM F 1506, *Specification for Textile Materials for Wearing Apparel for Use by Electrical Workers Exposed to Momentary Electric Arc and Related Thermal Hazards,* 1998.
10. Mary Capelli-Schellpfeffer, Raphael C. Lee, Mehmet Toner, and Kenneth R. Diller, "Correlation between Electrical Accident Parameters and Sustained Injury."
11. K. S. Crawford, D. G. Clark, and R. L. Doughty, "Motor Terminal Box Explosions Due to Faults."
12. This account is based on an actual incident. The names, including the name of the facility, have all been changed to protect those involved. Any similarity to actual names or facilities is strictly coincidental.

Answers: 1. (false), 2. (false), 3. (true), 4. (true)

■ Chapter 3

Hazard Analysis

Determining Exposure

A hazard analysis is the application of critical thinking to jobs or tasks that identifies *whether* or *how* a person might be exposed to hazards during job execution. The prime objective of the analysis is to avoid or minimize hazard exposure. The analysis should indicate the following:

- If exposure to a hazard exists
- The degree of exposure
- If personal protective equipment (PPE) is needed
- If some type of specific authorization or other control process is needed

In the same way that energy is required for work (in the physics sense) to be done, energy is required for an injury to occur. The source of energy may be mechanical, in the form of a rotating device; kinetic, in the form of either a fall or falling object; or thermal, in the form of heated components, very cold components, or a very hot plasma. Although this book concentrates on electrical energy, it is important to keep in mind the result of human interaction with any energy source. These other energy sources must also be considered.

To avoid an injury, a person must recognize the energy source and take necessary steps to avoid any interaction with that energy. Of course, hazard exposure alone does not result in an injury. For example, it is possible to work from an elevated platform and not fall. It is also obvious that degrees of exposure vary: A fall from a 10-foot-high platform is more likely to result in an injury than a fall from a 10-inch-high platform. As the degree of exposure increases, so does the chance of injury. The objective of the hazard analysis is to determine the degree of potential exposure to all hazards to which a person may be exposed as he or she performs an assigned job or task.

Thinking through the Process

The flow diagram in Figure 3–1 illustrates a hazard/risk analysis thought process. The diagram examines a potential injury that could result from working on an elevated platform. The figure is intended to illustrate the suggested thought process and that logical queries can be established to determine the degree of hazard and hazard exposure. Following this thought process can help protect people from injury.

The hazard/risk analysis illustrated in Appendix D of NFPA 70E, *Standard for Electrical Safety Requirements for Employee Workplaces,* 2000 edition, considers only currently recognized electrical hazards of shock and arc flash. Normally, it is not necessary to have a flow diagram for reference in order to analyze potential exposure. All types of hazards should be considered in the analysis. For example, injury from falls or

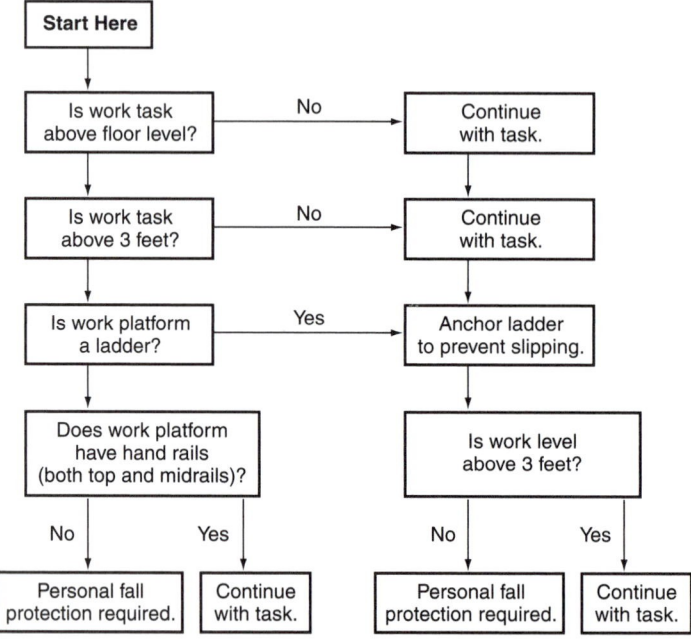

Figure 3–1. Personal Fall Protection Hazard/Risk Analysis Flow Chart.

falling objects must always be considered. The analysis may consist simply of one line of questions. Fall and falling object hazard analysis might also follow a similar pattern of thought. Of course, all types of hazards do not always exist. However, it is wise to consider all potential hazards when executing a hazard analysis in preparation for work.

When a job or task is repeated routinely, it might be possible to execute and document a hazard analysis as part of the standard procedure or practice. If the hazard analysis is documented in written form, the worker should review the analysis prior to starting work to make sure that there is nothing unusual about the situation.

The primary objective of the flow diagram in Figure 3–2 is to consider each electrical energy hazard and determine the degree of exposure. These are critical pieces of information if injury is to be avoided. From this data, it is possible to determine what personal protective equipment is warranted, as well as any procedural steps or controls that are deemed important.

> **NOTE:** Any task or job requiring exposure to shock should at least require that a "shutdown request" be executed. Production and process operation decisions should be made by supervisors responsible for such processes instead of supervisors responsible for electrical work processes.

The diagram in Figure 3–2 can help with the hazard analysis procedure. It is included to illustrate that the flow diagram concept shown relating to shock and arc flash can apply in a general procedure.

Using the Hazard/Risk Analysis Flow Chart

Block 1. In blocks 1, 1a, and 1b, the objective is to determine if an electrical hazard exists. General concensus suggests that 50 volts is the potential level at which a shock hazard begins. Although Figure 3–2, NFPA 70E, and 29 *CFR* 1910, Subpart S all suggest that voltages less than 50 volts are not electrically hazardous, it is important to understand that only shock and arc-flash hazards are considered. Other hazards do exist and should be considered. In many instances, programmable logic controller (PLC) and digital control system (DCS) conductors are energized below 50 volts, but still could cause a process interruption that

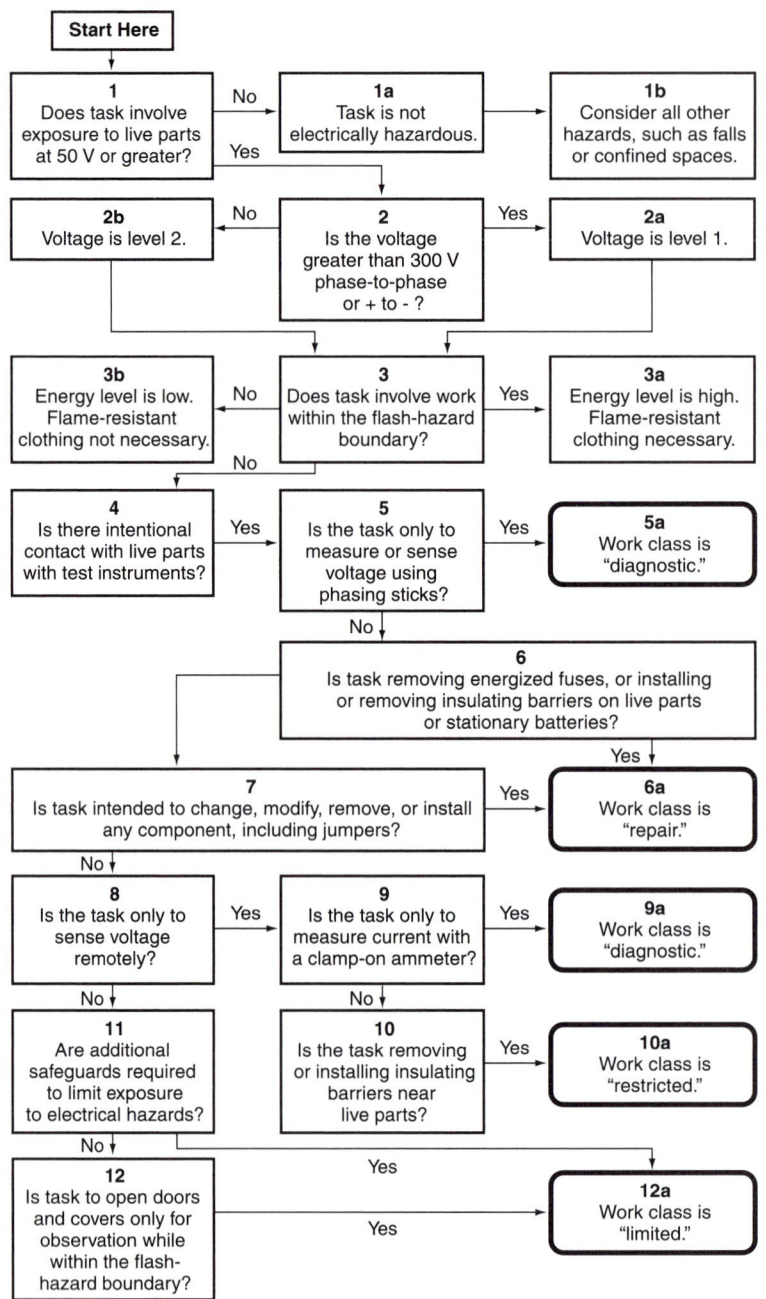

Figure 3–2. Hazard/Risk Analysis Evaluation Procedure Flow Chart.

Source: Modified from NFPA 70E, *Standard for Electrical Safety Requirements for Employee Workplaces.* Quincy, MA: National Fire Protection Association, 2000.

could result in an incident of major proportion. The flow diagram does not consider that batteries can store large quantities of energy with large arc-flash boundaries. It is important to understand what is meant by the answer to each question in the diagram. If a hazard exists, proceed to block 2.

Block 2. In blocks 2, 2a, and 2b, the objective is to determine the degree of shock hazard. As the voltage level increases, the *degree* of the shock hazard also increases. In other words, contact with 480 volts is more likely to result in an electrocution than contact with 115 volts. By understanding the degree of the hazard, it becomes possible to select an appropriate procedure, authorization, and/or protective equipment. Proceed to block 3.

Block 3. In blocks 3, 3a, and 3b, the objective is to determine the degree of arc-flash hazard. At this point in the thought process, the arc-flash boundary dimension is required (see Chapter 13, "Safe Work Practices"). Of course, as the quantity of available energy increases, the potential for injury from arc flashes also increases. With this information, again, appropriate authorization and/or flame-resistant clothing can be selected. Proceed to block 4.

Blocks 4 and 5. In blocks 4 and 5, the objective is to begin to determine the type of task being considered. Block 4 determines if intentional contact with live parts is being planned or considered. For example, in order to test for voltage, contact must be made. In block 5, more detail is sought about why intentional contact is anticipated. If the answer to this question is "no," the work must be diagnostic in nature. Therefore, the work category could be called "diagnostic."

Blocks 6 and 7. In blocks 6 and 7, the objective is to continue to learn more details about the intended task. A "yes" answer suggests that repair is under way. Of course, any equipment in need of repair should be deenergized. In some instances, a repair-type task might be needed in the vicinity of a different circuit. The authors strongly suggest that an electrically safe working condition be implemented prior to executing any repair work.

Blocks 8, 9, 10, and 11. The objective of these blocks is simply to gain more information about the intended task. Blocks 9a and 10a suggest that the degree of exposure to a person taking a current reading and removing barriers is unique. Appropriate category names are assigned to the work category.

Block 12. The question in block 12 is the last to be asked. By answering "no" to all preceding questions, the task must be only to look inside the door or cover. The objective of this block is to emphasize that simply opening a door for a peek is hazardous because energized conductors and terminals frequently exist inside the door. Of course, potential for an arc-flash event exists any time the door or cover is not completely secured.

Of course, after a work category is identified by executing a hazard analysis, the person should be able to select an appropriate procedure, authorization, and/or PPE.

Considering All the Hazards

Figure 3–2 considers only shock and arc flash. However, it is wise to consider all potential hazards when executing a hazard analysis in preparation for work. It may not be necessary for the hazard/risk analysis to be in written form. All recognized hazards should be considered in the analytical process.

When a job/task is repeated routinely, it is usually possible to write and maintain a written hazard analysis as a part of the standard procedure or practice. If the hazard analysis is documented in written form, the worker should review and concur with the analysis before starting work.

Each possible hazard should be considered in a similar manner. It is important to understand that the suggestion is not for a difficult or long analysis. Normally, only a few seconds are necessary to work through this thought process for each hazard. This type of analysis can consider all hazards in a few minutes. Experience has shown that taking the time to think about each hazard in this way improves the chance that nothing will be overlooked.

The general idea is to provide the best chance that a worker will not be injured as he or she executes the job/task. Sometimes, the hazard analysis indicates a need to wear PPE. After completing the initial hazard analysis, it is wise to consider how an item of PPE might be impacted by another hazard. Because PPE normally is designed and manufacturered to provide safeguard from a specific hazard, it is possible that exposure to a different hazard may degrade the protective characteristics of the equipment. For example, in some cases flame-resistant clothing is suggested for protection from arc flash. The same task might suggest a need for fall protection. It is important to understand what effect arc flash might have on the fall protection.

When performing an analysis of electrical hazards, it is extremely important to consider actual circuit parameters. In other words, any arc-flash analysis must be based on components as they are installed rather than on what is supposed to be installed. Frequently, installed overcurrent protection is not the same as intended. Early in the morning or late in the evening, when equipment is down and product is not being produced, a maintenance electrician's requirements might well differ from design requirements. He or she might choose a fuse or a circuit breaker based on what is available instead of what was specified. Overcurrent protection is the easiest and most likely component of a circuit to be different from the design specification. At the same time, although the overcurrent device has no bearing on the shock hazard, it has the greatest bearing on the amount of energy released by an arc flash.

 ## A Closer Look

When Len[1] left for work on Monday morning, he knew that he would be performing maintenance on unit substation 2A-8. The continuous-processing facility where he worked would be taking a short annual turnaround this week. Len would be working alone in the substation after it was locked out.

The process shutdown began at midnight on Sunday, with a plan to begin equipment lockout at 8:00 A.M. Knowing that the shutdown had been shortened from the normal two weeks to one week, management had brought in an outside contractor to supplement in-house maintenance personnel. However, no additional electrical workers were among the extra help.

Johnny was Len's supervisor. He and Len proceeded to substation 2A-8 to review Len's work assignment for that day, which was to do the following:

- Open and lock out each secondary breaker.
- Open and lock out the primary disconnect.
- Observe ammeter and voltmeters on the front of the switchgear.
- Open the meter compartment and remove the "range fuses" in the control circuit.

- Open one secondary circuit breaker and verify that no voltage exists on both the line and load side.
- Install locks on all secondary circuit breakers.
- Verify that the primary switch is open, test for a deenergized condition, and install locks on the primary.

Once these locks were in place, Len was to open all compartments and clean the inside of each one. Len would verify that cable terminations were tight and that overcurrent devices still met design criteria.

Johnny and Len stepped through the hazard/risk analysis. They reviewed the possibility for shock, then for arc flash. Class I gloves would be worn while checking voltage. Arc-flash equipment would be worn until all locks were installed. Len would be using an electric vacuum cleaner, so the cord would need to be inspected for damage, and he would use a portable ground-fault circuit interrupter (GFCI). All of Len's work would be done from the floor level, so the possibility of falls from elevation was not a problem. No other crafts would be working in the room, so co-occupancy was not a problem. The batteries in the rack would not be under charge, and their protective covers would remain in place; therefore, no problem with batteries existed. The line side of the primary switch was still energized; Len would not work in that compartment. No alternate or backfeeds would be involved with the work.

Both Len and Johnny had thought through the work and considered the hazards. Locks and tags were in place, and Len was ready to go to work. As Johnny left for another area of the plant, he told Len that he would check back with him in a couple of hours.

Len began his work by opening all the compartment doors to clean the inside of the equipment. After inspecting the vacuum cleaner cord, Len began to vacuum the equipment. Within an hour, he had completed cleaning in the secondary circuit-breaker compartments. He really wanted to clean at least the floor in the primary switch. Len thought, "I'll just clean the dead bugs and dust from the floor. I'll stay away from the line side. Anyway, it's hard to reach the line side." Len's judgment was impaired momentarily. He opened the primary compartment door, proceeded to clean the compartment, and closed the door. Len got away with a hazard exposure.

Next, Len opened the metering compartment door. He would clean the area, then check each point on the terminal block to make sure all were tight. The terminal block was located near the rear of the compartment. Len had to reach out in order to check the terminals. While he reached out, his arm rested against the lower edge of the compartment. The edge was not exactly jagged, but it was not exactly smooth, either. As Len tried to reach the terminal block, the compartment door swung closed, hitting Len in the back. His shirt was now wet from perspiration.

Mounted on the inside of the metering compartment door was a voltmeter that measured the space-heater voltage. Len had not considered the space heaters because his work did not require him to be near them. He also had not considered that the space-heater circuit might be connected to the voltmeter, but it was. Actually, he was not very concerned about the 110-volt hazard, anyway.

When the door swung into Len's back, a terminal on the voltmeter touched him, producing a small shock. He was very surprised and instinctively reacted. The movement caused his arm to drag across the rough edge of the compartment, which sliced it open. His arm was bleeding badly. A vein must have been cut, from the amount of blood that was streaming from the arm.

Len wrapped his shirt around the wound and headed toward first aid. The flow of blood had slowed, but only slightly. After he left the substation room, another worker saw the problem and called for help. The coworker helped to apply pressure to the wound until the company nurse arrived. Len was transported to the nearest medical facility, where he received adequate attention in the emergency room.

Neither Len nor Johnny had considered this type of injury. Both had been very concerned about the transformer, the big cables, and the big bus structure. Neither had considered metering equipment mounted on the door.

Test Your Thinking

1. A hazard analysis should indicate the following:
   a. If exposure to a hazard exists
   b. The degree of exposure
   c. If PPE is needed
   d. If authorization or other control process is needed for the job
   e. All of the above
2. Production and process operation decisions should be made by the following:
   a. Supervisors responsible for production and process operations
   b. Supervisors responsible for electrical work processes

3. To avoid an injury, a person must do the following:
   a. Recognize the energy source
   b. Take necessary steps to avoid interaction with the energy
   c. Both (a) and (b)
4. Which of the following conditions must exist for an injury to occur?
   a. A hazard
   b. Exposure of a person to a hazard
   c. Contact with an energized circuit
   d. A release of energy (accident)
   e. Choices (a), (b), and (d), but not (c)

Note

1. This account is based on an actual incident. The names, including the name of the facility, have all been changed to protect those involved. Any similarity to actual names or facilities is strictly coincidental.

Chapter 4

Exposure Management

Risk Management and Personal Exposure to Hazards

The existence of a hazard does not necessarily result in exposure of people. No risk of personal injury exists if no exposure exists.

It is widely known that three conditions must exist simultaneously for a fire to occur: Fuel, oxygen, and an ignition source must all be present (see Figure 4–1). Eliminating any "leg" of the fire triangle eliminates the possibility of a fire.

Similarly, three conditions must exist for an injury to occur: A hazard must exist, a person must be exposed, and a release of energy (accident) must simultaneously occur. Eliminating any "leg" of the injury triangle eliminates the possibility of an injury (see Figure 4–2).

The elements of the fire triangle relate to physical components. The elements of the injury triangle relate to soft human components.

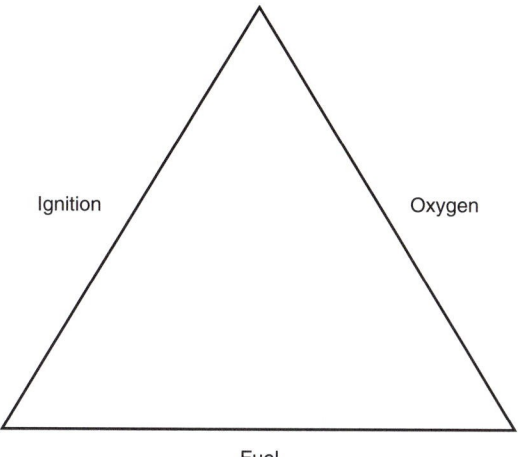

Figure 4–1. The Fire Triangle.

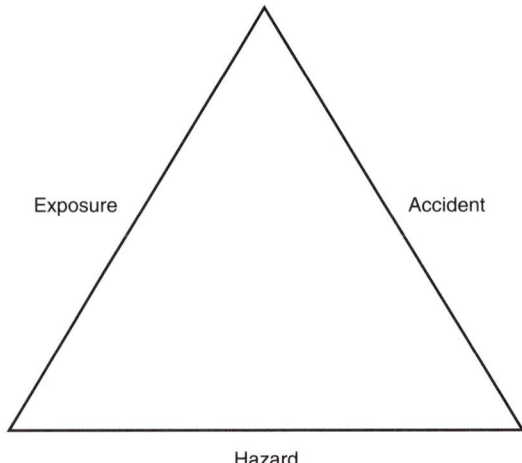

Figure 4-2. The Injury Triangle.

The Injury Triangle

An electrical safety program should be designed and implemented with the injury triangle in mind. Each component of the program should attempt to influence only one leg of the triangle. Just as fables told through the years contain a single message or moral, a second objective would result in a less effective component. It is much better for the secondary objective to be addressed in a separate program element. For instance, a training program element might provide information about electrical hazards. That element should focus the attention of both the leader and the students into a discussion of a single hazard. Of course, for an electrical safety program, the "hazard" leg means discussing the electrical hazards identified in Chapter 2.

Getting the maximum safety benefit from each program element of the electrical safety program is best achieved when the elements are directly associated with a leg of the injury triangle.

The "Hazard" Leg

The *National Electric Code*® and the *National Electrical Safety Code* (*NESC*) address issues related to the hazard leg of the injury triangle. These standards cover enclosures, overcurrent protection, wiring integrity, grounding (earthing), and similar issues. The general idea is to avoid the existence of a hazard.

NFPA 70E, *Standard for Electrical Safety Requirements for Employee Workplaces,* 2000 edition, and OSHA Subpart S cover issues related to work practices. NFPA 70B, *Recommended Practice for Electrical Equipment Maintenance,* 1998 edition, covers issues related to maintaining equipment integrity. An accident is almost always associated with a person acting in an inappropriate way (NFPA 70E) or with an equipment failure (NFPA 70B). The general idea is to avoid any release of energy, as illustrated by the accident leg of the injury triangle.

Some consensus standards contain a training requirement. Standards use the term *qualified employee* or a term with similar connotation to highlight the ability of a person to recognize an electrical hazard. Recognizing an electrical hazard is similar to the idea of recognizing quality: "I'll know it when I see it." Electrical hazards normally are not directly apparent. Only by seeing and reacting to visible indications can an electrical hazard be recognized.

The "Exposure" Leg

This leg of the injury triangle is not fully covered by any consensus standard. Bits and pieces of some standards cover some aspects of exposure. The concept of eliminating exposure tends to be lost in the depths of other aspects of the program.

Like consensus standards, electrical safety programs often do not directly address how to manage exposure. A tour of a plant or an electrical control room of a shopping center will likely find doors ajar, covers held in place with two or three bolts, and unused holes in the top and sides of electrical equipment—all practices that increase exposure to hazards. It is also likely that a coffeepot or assembly desk is not only in the room with but also in front of electrical equipment, thus causing people to increase their own risk exposure. The electrical safety program should include elements that discourage people from gathering in the vicinity of electrical equipment and encourage employees to close all fasteners.

Exposed energized components generally are accessible only within the confines of a fence, vault, or other guarded area. Although equipment failure could result in significant equipment damage, no potential for injury exists unless a person is nearby (exposure). Even if a person is near, no injury occurs unless an accident (release of energy) occurs.

If a person enters a vault, substation fence, or enclosure, the potential for injury becomes a reality. Respecting both shock and arc-flash boundaries is one way to avoid exposure.

The "Accident" Leg

All safety initiatives target accident reduction as an objective. It is generally accepted that unsafe conditions, unsafe practices, and unsafe equipment cause all accidents. These ideas make up the accident leg of the triangle. Many consensus electrical standards such as those from Underwriters Laboratories (UL) and the National Electrical Manufacturers Association (NEMA) attempt to address the problem of unsafe equipment. NFPA 70B also attempts to reduce the exposure to unsafe equipment through maintenance. Installation standards, such as the *NEC*, attempt to influence unsafe conditions. However, each of these standards deals with equipment operating "normally." During times of "not-normal" operation, the issue boils down to work practices and accidents. For all practical purposes, the accident leg of the triangle is based on work practices, and inadequate work practices lead to the vast majority of injuries. The electrical safety program should consider all elements of the accident leg but place the majority of its emphasis on work practices.

It is generally accepted, and the authors concur, that electrical facilities, services, and distribution lines that are designed and installed to meet the requirements of the *NEC* and the *NESC* are safe when operating normally. However, should the installation or equipment enclosure be abridged for any reason, all bets are off. Neither the *NEC* nor the *NESC* address the "not-normal" set of circumstances. Both of these standards include requirements intended to provide for safety when conditions are not normal, but neither contains the "people element."

The Maintenance Program

Equipment and services begin to deteriorate from the moment of installation. As the deterioration increases, workers' exposure to an electrical hazard increases. One of the objectives of the maintenance element of a safety program, then, must be to maintain the electrical installation close to the original installed condition. A maintenance program should be implemented that ensures grounding integrity and insulation integrity.

Although intentional grounding is always supposed to limit any potential (voltage) difference between two points, there are additional objectives. If the potential difference between any two points within reach by a person to a voltage can be limited, which does not permit a lethal current to flow, that source of electrocution can be avoided. Initial consensus requirements accomplish this purpose. The maintenance program must provide increased assurance that the initial objective continues through the life of the equipment.

In general, the grounding system relies on the earth to establish "zero volts" as a frame of reference. All other components, then, can be judged against that reference point.

The maintenance program should be a written document of standard practices or procedures. Continual program review and adjustment should be performed as necessary to keep the program current. For instance, testing ground rods every year is unnecessary if no problems are found. Instead, the maintenance testing frequency should be adjusted until a need is identified.

Generally, programs adequately cover important cleaning and visual aspects of maintenance. However, some programs inadequately consider circuit components. For example, many enclosures are modified to add new components or to replace existing ones with upgraded devices. As equipment enclosures are modified, the initial integrity of the enclosure is compromised, resulting in a potential projectile in the event of a rapid increase in pressure within the enclosure.

Choice of Purchased Equipment

Generally, instructions provided by manufacturers adequately address the shock hazard. Yet few instructions, if any, include information related to arc-flash or arc-blast exposures. The purchaser must consider arc flash and blast in the design of the maintenance program. Some purchased equipment might contain specific requirements or components that introduce the possibility of arc flash. For example, electrical equipment usually generates heat and requires ventilation. In almost all instances, the manufacturer provides ventilation screens or vents on the front of the equipment to dissipate the heat. These vents or screens are excellent pressure-relief mechanisms. The problem comes when a person is present, as he or she is likely to be standing in front of the vent or screen; thus, the person is exposed to hazard in the event of an arcing fault. A much better approach would be to avoid, by equipment design, any potential energy release from the front of the equipment.

As demands for electrical energy increase, a trend has developed toward installing larger unit substations with smaller internal impedance. The discussion in Chapter 2 of arc-flash hazard points out that circuit impedance is one major factor in the determination of the arc-flash boundary. Much of the impedance of a circuit is in the transformer supplying the circuit. As the internal impedance decreases, available short-circuit current increases, with a resulting increase in the arc-flash boundary. Internal impedance should not be held artificially low.

The arc-flash boundary (within which people are exposed to arc flash) is also heavily dependent on the speed of the overcurrent device (see Chapter 13, "Safe Work Practices"). Because energy is the damaging culprit in arc-flash events, rapidly clearing the fault is very important. In most instances, overcurrent protection is selected that will protect the equipment or circuit from destruction. To manage risk of injury from arc flash effectively, designers and engineers should select overcurrent protection that minimizes potential exposure to arc flash. In other words, select overcurrent protection based on protection of people instead of protection of equipment.

Area Classification

One generally accepted process of managing risk is the system of designating a classification based on the possibility of a fuel explosion from an electrical source occurring in the area (see NFPA 497, *Recommended Practice for the Classification of Flammable Liquids, Gases, or Vapors and of Hazardous (Classified) Locations for Electrical Installations in Chemical Process Areas,* 1997 edition. The risk of an explosion is never eliminated, but it can be reduced to an acceptable level.

For a fuel explosion to occur, all three elements of the fire triangle must be present. Several protection plans intend to eliminate one leg of the fire triangle. For example, equipment purged with clean air eliminates the possibility of ignitable material accumulation. In other plans, the fuel leg of the fire triangle is interrupted. In turn, the hazard leg of the injury triangle is also interrupted.

Location of Equipment

Engineers and designers normally locate equipment based on economic considerations. Motor control centers (MCC) and power distribution panels (PDP) frequently are located close to the user equipment in order to reduce the length of the wire. If a person is near a PDP or MCC when an arcing fault occurs, an injury is likely. It is not unusual for a door or cover to be less than completely latched. In this instance, the degree of exposure is elevated.

The authors strongly recommend that all equipment serving to distribute electrical energy be located so that it is physically remote from "normal" passageways. Electrical control rooms (ECR) or areas with electrical equipment should remain free of coffeepots, bulletin boards, or other inducements for people to gather.

Effective Communication Systems

A site or organization should have an effective system to enable complete and direct communication between and among work teams and crews who are working on different parts of the same system or systems that are interlocked with each other. The key value of effective communication is that employees and work teams or crews know the boundaries of their responsibility. Employees should be able to communicate how their work affects other workers or work teams. Effective communication enables effective planning that, in turn, decreases exposure.

The ability to manage risk is heavily dependent upon the intent of the organization's policy. More than that, the real intent of the organization's management is visible to employees, and employees tend to implement or execute a policy as they perceive it to be. Discussions with employees that include questions and directions related to the safety policy suggest that supervisors emphasize avoiding exposure to hazards. Employees usually interpret discussions with employers that cover only schedule, cost, and physical elements of a task or job to mean that avoiding exposure to hazards is not necessarily important. An effective communication system involves all communications with employees.

🔍 A Closer Look

When the Pulaski Cellulose[1] plant was built several years ago, a spare circuit breaker was installed in preparation for future expansion. The cellulose plant produced a product for a few years before the planned expansion was necessary.

A project planned to use the spare breaker for a new distribution panel service. Because the circuit breaker had been used as a spare for other breakers several times, the electrician removed the circuit breaker and returned it to the original manufacturer to verify the condition of the breaker. The work was done, and the breaker was returned and inserted into the compartment. At the Pulaski plant, the secondary switchgear construction was three units high, with this "new service" in the top-most position. The compartment door had vents installed to prevent any water from entering, even though this particular switchgear was designed for installation indoors, on the ground floor of a multistory building.

When the circuit breaker was closed, something (perhaps a short piece of wire) fell from a ledge at the top of the compartment and initiated an arcing fault on the line side of the "new" circuit breaker. In order to reach the operating handle of the circuit breaker, the electrician was standing directly in the path of the vents in the door. Plasma and gases were expelled through the door vents directly into the face of the electrician. The electrician received third-degree burns over his upper torso and spent several weeks in the hospital.

Although water egress is a significant problem, the vents could easily have pointed up instead of down. No water could have possibly entered the vents, because the switchgear was located in the building. Even if it had, the water would simply have run down the inside of the switchgear and onto the floor. Arc flash and blast are frequently major hazards, and the existence and direction of vents that provide pressure relief must be considered in order to manage risk associated with arc flash and blast.

Test Your Thinking

True   False

☐   ☐   1. If a facility is designed and installed to meet the requirements of the *NEC* and the *NESC*, the facility is safe from electrical hazards when operating normally.

☐   ☐   2. The grounding system relies on the earth to establish "zero volts" as a frame of reference.

☐   ☐   3. The arc-flash boundary has nothing to do with the speed of the overcurrent device.

☐   ☐   4. All equipment serving to distribute electrical energy should be located so that it is physically remote from normal passageways.

Note

1. This account is based on an actual incident. The names, including the name of the facility, have all been changed to protect those involved. Any similarity to actual names or facilities is strictly coincidental.

Chapter 5

Strategies for Preventing Injury

Understanding the Current State of Knowledge

In the United States, injuries are tracked in several different ways. The medical community maintains records of injuries and their treatment so that all practitioners—whether across town or across the country—can access the same information about the most successful treatments. Generally, the Centers for Disease Control and Prevention (CDC) maintain significant records. Those records are categorized only by the type of injury. For example, a burn injury is recorded as a burn, and little information is recorded related to the cause of the injury. In some cases, a burn may be recorded as an electrical burn. Such a record provides better information but still leaves the question unanswered about the source of the burn. (For more information, see Chapter 16, "The Changing State of Electrical Safety.")

Injury records are maintained in other communities, as well. In the United States, OSHA requires an employer to maintain a record called the *OSHA Form 200* (see Figure 5–1). This record is intended to document employee illnesses and injuries experienced. That record, then, provides information on where the employer should place its resources to achieve improvement. However, the record still does not delve into the type of energy related to the cause of the injury.

Injury records are important because they can provide information to prevent similar injuries. However, the record frequently does not provide enough information to allow corrective action. CDC records a burn injury as a burn, with no mention of the energy source. A similar *OSHA Form 200* could also record a burn. The injury might even be recorded as an electrical burn. However, it is unlikely that the record

U.S. Department of Labor

For Calendar Year _____   Page: _____ of _____

Company Name	Form Approved
Establishment Name	O.M.B. No. 1218-0176
Establishment Address	See OMB Disclosure
	Statement on reverse.

Extent of and Outcome of Injury

Type, Extent of, and Outcome of Illness

Fatalities	Nonfatal Injuries					Type of Illness							Fatalities	Nonfatal Illnesses				
Injury Related	Injuries with Lost Workdays				Injuries Without Lost Workdays	CHECK Only One Column for Each Illness (See other side of form for terminations or permanent transfers)							Illness Related	Illnesses with Lost Workdays				Illnesses without Lost Workdays
Enter Date of death. mm/dd/yy	Enter a Check if injury involves DAYS away from work or restricted work activity or both.	Enter a Check if injury involves DAYS away from work.	Enter number of DAYS away from work	Enter number of DAYS of restricted work activity	Enter a Check if no entry was made in column 1 or 2 but the injury is recordable as defined above.	Occupational Skin Disorder or Disease	Dust Disease of the lungs	Respiratory Conditions due to toxic agents	Poisoning (systemic effects of toxic materials)	Disorders due to physical agents	Disorders associated with repeated trauma	All other occupational illnesses	Enter DATE of death, mm/dd/yy	Enter a CHECK if Illness involves DAYS away from work, or DAYS of restricted work activity or both.	Enter a CHECK if Illness involves DAYS away from work.	Enter number of DAYS away from work.	Enter number of DAYS of restricted work activity	Enter a CHECK if no entry was made in columns 8 or 9
(1)	(2)	(3)	(4)	(5)	(6)	(a)	(b)	(c)	(d)	(e)	(f)	(g)	(8)	(9)	(10)	(11)	(12)	(13)

(7)

Certification of Annual Summary Totals by: _____   Title: _____   Date: _____

POST ONLY THIS PORTION OF THE LAST PAGE NO LATER THAN FEBRUARY 1

Figure 5–1. *OSHA Form 200.*

would differentiate between an arc-flash burn and a burn from current flow through tissue.

All injuries are the result of human interaction with energy. The energy release might be the result of either intentional or accidental action. The action might be the result of an equipment failure or human action, but energy is always involved. For example, an automobile accident involves momentum (kinetic energy). Without the momentum, the vehicle would not be moving; therefore, an injury could not happen.

To intervene and prevent incidents and the resulting injuries, it is necessary to modify, limit, or otherwise impede the interaction of a person with the energy. In a gasoline fire, a chemical transformation results in the release of thermal energy. To prevent the fire, the chemical transformation must be prevented. Once the fire has started, the intervention must limit exposure to the resulting thermal energy. Providing a barrier between the person and the thermal energy prevents injury. The barrier can be protective clothing, or it can simply be physical space.

Records generally do not provide sufficient information to determine the type of energy after the circumstances have been forgotten. Even after reports became commonplace, the injury records make no distinction between the source of a burn or an electrical burn, effectively hiding the existence of an arc-flash hazard. It is also quite possible that other electrical hazards remain hidden.

Electrical shock has been known to be a hazard since the early days of electrical energy use. Categorizing and recording noncontact burns simply as "burns" for many years resulted in arc flash being masked. Even after the publication of "The Other Electrical Hazard" in the mid-1980s, the arc-flash hazard was only recognized as a hazard by a consensus document with the publication in 1995 of NFPA 70E, *Standard for Electrical Safety Requirements for Employee Workplaces*. It is not clear that arc flash is mentioned in any OSHA standard.

Although not required by any generally accepted document, an effective historical injury record should contain information related to the type of energy as well as the type of hazard causing the injury. In fact, the record should contain sufficient information to allow an investigator to reconstruct the essential elements of how the injury occurred. Only with sufficient information are historical records effective in guiding an injury intervention process.

It is not that the records currently required are ineffective. Indeed, these records substantiate the existence of a broad safety problem and justify insurance rates. But supplementary information would help to identify an intervention process and aid injury prevention. A hazard must be identified as an injury source before any practical means to avoid injury can be identified.

In an arc-flash incident, a significant pressure wave is generated (see Figure 5–2). That pressure can result in a motor-terminal box exploding or a compartment door opening during the fault. How the human body reacts to the pressure wave is not yet known. Researchers at the Electrical Trauma Center of the University of Chicago report in IEEE papers that injury is possible, and even likely, from these large pressure-wave fronts.[1,2]

An arc flash is always accompanied by a plasma, which has been described by observers as a "ball of fire." The plasma is thermally very hot, as discussed in Chapter 2. The plasma is an ionized gas that is conductive—indeed, is carrying current. When current is flowing, a magnetic field exists. It is frequently reported that a person exposed to an arc-flash event is surrounded by a ball of fire. Parts of the body, or even the entire body, are within the conducting plasma. No one knows what happens to human tissue exposed to these conditions. Essentially, no research has documented how tissue (particularly brain and nerve tissue) reacts during or after exposure to magnetic plasma. Injury records could easily mask this type of exposure.

Employers and industry associations do not have to wait for regulatory action or a national consensus standard to define the needs associated with records. Either a single employer or an association of employers could begin to maintain supplementary records.

Employers generally are reluctant to share injury records, especially in the litigious climate that exists in the United States. Industry associations and employers could establish complete records of injuries, even incidents, where an injury is not involved. Employers generally believe that fewer records are better. From a litigation point of view, that view most likely is accurate. However, more complete records would result in an improved chance of preventing injury in the first place. By reducing injuries, an employer or an association could also reduce injury-related costs. Although records alone might offer little help, without them, no understanding is possible.

Figure 5–2. Photos of an Arc Flash.

The People Factor

As suggested in Chapter 2, most electrical safety initiatives have been associated with equipment construction and installation. At the same time, most incidents are initiated by *people*. In the majority of cases, a person exercises poor judgment and takes an action that should not have been taken. It is even possible, but very unlikely, that the poor judgment is intentional.

The ability to exercise good judgment is based on a few foundation factors: experience, skill, and knowledge (see Figure 5–3). Even after a person achieves the necessary knowledge, skill, and experience, effective judgment is not automatic. These factors provide the basis for executing the judgment. Training programs usually provide knowledge information and some form of skill training. However, the only way to gain experience is through practice.

Good judgment is the application of knowledge, skill, and experience in a way that enhances the decision-making process, resulting in good decisions. Individual thought patterns affect the way that a person applies his or her knowledge, skill, and experience. To exercise good judgment, a person must avoid thoughts that affect judgment in a negative way. Essentially these thoughts fall into five types of thinking, identified as follows.

Figure 5–3. The Foundation of Judgment Triangle.

1. **"Don't tell me! I helped write the rules, so I know what I'm doing."**
   This is the first type of thinking to be avoided. This thought is exhibited by a person who resists following procedures, plans, or rules. This thinking should be replaced with the thought: *"Follow the procedure. I helped write it, so it must be good. I might forget a step or get a step out of sequence if I don't use the procedure."*
2. **"Do something fast. The process is down."** This type of thought is usually an honest attempt to avoid an unpleasant situation. The person is reacting sincerely in an attempt to act quickly; however, the thought often results in impulsive action, and impulsive actions can result in initiating an incident. This reactive reasoning must be replaced with the thought: *"Slow down, think, and don't act impulsively."*
3. **"It won't happen to me. I've done that for years, and nothing has happened yet."** Unfortunately, this type of thought is common among electricians and electrical engineers. It is important to remember that electricity is patient. It will wait for years. Then, when the moment is right, it reaches out and bites. This type of thinking must be replaced with the thought: "It **could** *happen to me, and I could be the next statistic."*
4. **"Of course I can do it."** This type of thinking occurs when a person feels that he or she is well trained and highly skilled. Even though that may be true, it is wise to remain cautious and obtain feedback or direction before proceeding. This overconfident thinking must be replaced with the thought: *"Everyone knows that I'm good. I need to protect that opinion and show them that I don't take unnecessary risks."*
5. **"What's the use?"** This type of thought is almost the opposite of that of the overconfident person. Indicative of a person who doesn't really expect to develop the necessary skill and understanding, this thought pattern frequently results in an action that should not have been taken. This underconfident thinking must be replaced with the thought: *"I will not take any unnecessary risk. I can show that I do know what I'm doing."*

Training

Each training program must have a clear objective. To improve the electrical safety program, that objective must improve the student's ability to execute good judgment. Therefore, the training program should address each element illustrated in Figure 5–3. Training sessions can address either a discrete element or a combination of more than one element. However, the end result of the training should be to enhance the student's judgment ability.

In most instances, knowledge and skill are frequently the first, second, and *only* elements considered for training. Typically, training programs produced and provided by equipment manufacturers cover their particular equipment or product. But in some instances, training outlines offered by a manufacturer provide excellent information about an entire class of products. The objective of these program outlines is to provide the student with specific knowledge about the product or product class. Although employers and employees should take maximum advantage of these programs, they should also understand the limitations.

Only in rare instances do training programs provided by a manufacturer contain information related to safe work practices. They also typically do not consider the people factor or the work environment. These programs tend to address only the knowledge foundation element. This does not mean that the programs are not effective, but that they have a limit or boundary. The user must understand the manufacturer's motivation behind the program's production: Although most equipment manufacturers care about safety to a high degree, the manufacturer wants to sell its *own* product—not some other manufacturer's product. One notable exception to a profit motive is Cooper-Bussmann's "Safety Basics" program, which illustrates one manufacturer's concern for the people factor.

Apprentice programs provide effective training. In most instances, an apprenticeship is intended to help build or enhance a person's skill. In order to increase skill, knowledge must first be accumulated. The skill, then, is the application of that knowledge to accomplish a task. *Skill* suggests sufficient knowledge for a person to select and effectively use appropriate tools to accomplish a task requiring physical ability and dexterity. Developing skill requires on-the-job training (OJT) because skill is associated with executing a physical task. Although some classroom training may be required, the only real option is to execute the task on the job site.

Like knowledge training, skill training has limitations and boundaries. Electrical safety issues arise while experience is being gained during the skill training. In fact, the teacher is frequently the "old hand" who has accumulated many years of experience based on ingrained electrical safety training from many years in the past. It is likely that the apprenticeship is so focused on acquiring the necessary physical ability and dexterity that electrical safety issues are not covered. Safe behavior may not be the strong point of the "teacher." Herein lies the most important limitation of OJT: The student is taught to emulate the behaviors of the teacher.

Work practices are learned, and work practice training must be included as a significant part of skill training. A training program should be developed that is based on procedures, principles, values, and the work environment that exists in the workplace. In most instances, available work practice training programs suggest work practices that are deemed, by *someone,* to be good. Several excellent training programs are commercially available. Again, the limit or boundary of any commercially available training program should be understood. In most instances, these types of training programs usually do not address some important issues, such as policies, procedures, work environment, personal values, and principles. An employer must supplement any commercially purchased program to cover these elements.

Once skill and knowledge have been acquired, the student then begins to accumulate experience. As experience is gained, training takes on a different aspect, and concepts learned during training are reinforced. Corporate policies and values also play an important role as employees gain work experience. It is extremely important that electrical safety issues receive continuous reinforcement. Emphasis should be placed on procedures and work practices. Although many programs do not extend into the experience phase of training, coworkers, colleagues, and supervisors will reinforce some element of the previous training. This makes it important to ensure that they reinforce safe behavior and not unsafe behavior.

Knowledge Training

Basic technical knowledge must be a priority for every worker. In most cases, this type of information is best covered in a classroom setting. In general, the training should include an element of testing to measure the degree of knowledge that is gained in the program. Some consensus stan-

☑

- ☐ The effects of current flow in human tissue
- ☐ The effects of human tissue exposure to arc flashes
- ☐ Impedance of human tissue
- ☐ Impedance of human contact
- ☐ Concept of approach boundaries
- ☐ Flashover distances at various voltages
- ☐ Concept of flash-protection boundary, including how to calculate
- ☐ Hazards associated with testing circuits
- ☐ Effective construction of safety grounds
- ☐ Hazards associated with grounding
- ☐ Importance of communications
- ☐ How to listen
- ☐ Effects of pressure on an enclosure
- ☐ Care and inspection of a voltmeter
- ☐ Effects of voltage on current flow
- ☐ Effects of voltage on arc flash
- ☐ Unknown electrical hazards
- ☐ Relationship of exposure to hazards and injury
- ☐ Protective characteristics of PPE
- ☐ Construction and operation of electrical equipment (*Note: manufacturers training program usually covers this information*)
- ☐ Visual indications of an electrical hazard
- ☐ Different types of electrical hazards
- ☐ Existence and content of site procedures
- ☐ Existence and content of employer policies

Figure 5–4. Checklist for Knowledge Training.

dards (including OSHA standards) define a need for training records. The authors recommend that employee personnel files include a record of all satisfactorily completed training. In addition, outlines of the content of those training programs should be recorded and maintained.

The checklist in Figure 5–4 contains items that should always be included in electrical safety knowledge training. The checklist may be used to ensure that the items are included in site knowledge training.

Skill Training

Skill training should involve the application of each element of knowledge training listed in Figure 5–4. In most instances, skill training requires a combination of classroom and OJT. Skill training should be based on the

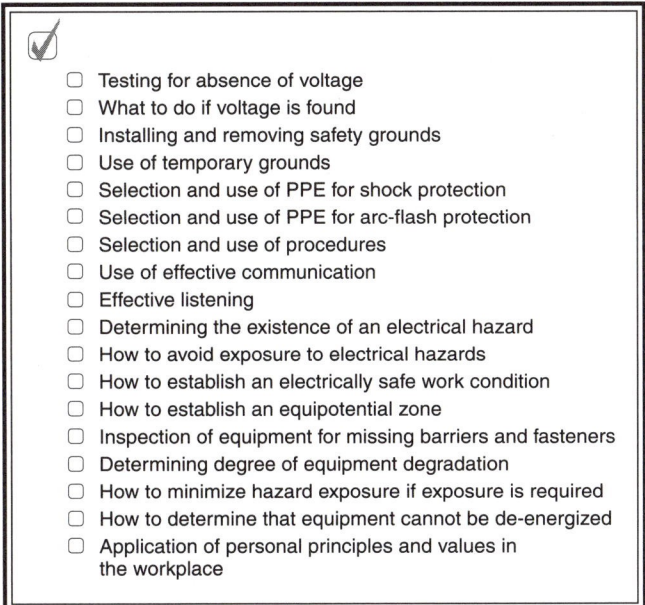

Figure 5–5. Checklist for Skill Training.

role that an employee is expected to execute. In other words, training should be designed and offered to an employee that is appropriate for his or her job. For example, technicians who are not exposed to 600-volt circuits do not need to have skills necessary to work on or near those circuits.

Electrical safety skill training should include information related to the elements found in the checklist in Figure 5–5.

Good Judgment Training

Training programs should always contain examples of real-life incidents. The incident scenario should be described in enough detail for the student to mentally understand and visualize the environment surrounding the incident. Usually when a person makes a mistake, the lesson learned is one that stays with the person for a long time. This is especially true if a personal injury resulted from the error. If a colleague was injured or narrowly escaped injury from an incident, an employee is likely to remember that point, as well.

Real-life incidents provide a fertile source for discussion of judgment. The way that the incident is discussed is important. If colleagues and coworkers were involved, the training must avoid emphasizing personal weaknesses in those involved. Instead, the discussion should emphasize judgment opportunities that could have been different. If the incident is not directly associated with employees at the site but is similar to circumstances there, the training could emphasize how the person(s) involved in the incident made errors in judgment.

The basic idea is that students develop experience in applying good judgment to incidents. Judgment training does not have to be formal. This training could be as simple as a detailed review of an incident reported in the newspaper discussed with workers at lunchtime. Continuing to place a high priority and emphasis in applying good judgment in incident scenarios must exist.

Work Practice Training

In many instances, site procedures define work practices and step-by-step directions of how a task must be performed. Some standards define acceptable work practices. However, work practices are really learned by observing an "old hand" as he or she performs a similar task. Routine discussions "in the shop" or "at lunch" generally weaken sound work practices. Those discussions tend to identify ways that productivity can be improved or shortcuts that can make the job easier to do. Generally, these discussions are not related to decreased exposure to hazards.

Site procedures should be formally reviewed from time to time. That review process should focus on a single procedure and avoid getting sidetracked by secondary issues. At the conclusion of each procedure review, each employee should understand what is required to comply with the requirements of the procedure. It is also important that the review include sufficient discussion so that each employee understands the reasoning behind each requirement. In addition, each employee should feel comfortable enough to challenge each requirement.

Work practice training must be a continuing effort. Observation and advice are required as tasks are executed.

Personal Principles

In many instances, a person is not immediately aware of the personal principles or values he or she holds. Although a person may act or react in a manner similar to another person, the action or reaction is frequently without conscious thought about what is the right thing to do. This experience is more likely in the workplace than in other settings.

Safety programs and initiatives should rely on a person's inherent value system. An effective safety effort helps the person to understand how his or her personal principles and values can apply in the work environment. It is unlikely that a person would consciously create a hazardous condition where either the worker or coworker is exposed.

Outside the work environment, an emotional connection exists between or among people. It is interesting to note that the *kind* of emotion is relatively unimportant. Values and personal principles are engaged effectively with this emotional tie.

Within the work environment, worker actions generally are guided by past experience and expectations of supervisors. Values and principles are not readily engaged within the worker's thought processes.

However, it is possible for the safety program to trigger the mental connection within employees. Principles and values should be discussed in a rational and nonjudgmental way. Discussions that engage an emotional reaction in the listener/student seem best able to trigger thought about values and principles.

An organization might avoid incidents and injuries effectively without rules and regulations. In an ideal environment, employees would have sufficient knowledge, skill, and experience to avoid all exposure to injury. In such an environment, each employee's judgment would be clear and guided by complete understanding, and his or her principles and values would provide all the necessary guidance.

Of course, the ideal environment is unlikely to exist in the workplace. Procedures and rules are needed. However, a principle-based effort relies increasingly on principles to guide actions in lieu of rigid rules, standards, and procedures.

A Closer Look

Ken Woods was a 47-year-old electrician at the Danville textile fiber plant.[3] His early career varied from a production operator at the plant, to a general mechanic, to an electrician. For the past nine years, Ken had been assigned as an electronic technician in the maintenance department. He was good at the task that he performed most often: to keep the solid-state drives operating at peak efficiency. The equipment manufacturer provided first-class maintenance training, and Ken seemed to have a special talent for this kind of work.

Ken had received adequate safety training, and he understood the safety procedures at the Danville plant. He always followed those procedures and had never been involved in an incident when someone was injured. In fact, the Danville plant had not had a lost-time injury in the past 14 years. The safety program was solid and effective.

This particular week, the Danville plant was completely shut down for an expansion to double the facility's capacity. One new 15-kV feeder with five new power distribution substations was being added. Management agreed to take a plant shutdown that would last 24 hours to make the cable terminations. The power distribution crew did not have enough people to expeditiously transfer loads that were not associated with the new feeder. (Administration and shipping were going to continue to operate, and some load transfer had to be done.)

Because Ken would not be needed in his normal job, he was temporarily assigned as an electrician to help the power distribution crew prepare the distribution system for lockout. The supervisor of the power distribution crew did not expect anyone assigned as a temporary helper to understand the distribution system. All load transfers and switch operation would be done by regular crew members. Temporary helpers would be assigned to work after the switches were opened.

Ken accompanied the supervisor and electricians as switch 22-4A was opened. He watched the arc grow as the overhead switch contacts moved to the open position. He watched the arc as it was being quenched and as it disappeared. Ken accompanied the supervisor to the nearest substation to observe the voltmeter on the face of the substation. The voltmeter read "0" volts. The line was deenergized.

Ken was assigned to install a set of grounds on the deenergized segment of the line. He was to climb onto the short tower that provided access to the switch. The load side of the switch was within reach from the platform. Ken noted the tag attached to the ground set. The tag indicated the ground-set number. The Danville standard practice was to track the location that ground sets are installed.

Ken accepted the assignment. The ground set was a little larger than he normally used, but he understood what was required. The supervisor reminded Ken that after the ground set was installed, the switch should be locked out. The supervisor then proceeded to the next location to continue execution of the planned work.

Ken placed the ground set and his voltmeter in the lift bag and climbed onto the platform. He pulled the canvas lift bag onto the platform and removed the ground set and tools. He looked up and observed the knife blades on the switch still open. He reached for the ground set and remembered that lockout procedure required a test for absence of voltage before any physical contact. Ken then left the ground set on the platform and reached for his voltmeter. The procedure required that the voltmeter be tested on a known source before using it to test for absence of voltage. With the line side of the switch still energized, Ken placed one probe on the grounded downcomer and reached toward the line side of the switch.

The arc started before contact was completed. Ken was trying to test for absence of voltage with the digital voltmeter that he always used in the electronics world. The meter literally exploded in his hands. Blast forces generated by the arc blew him across the platform and over the handrail. All at one time, Ken had been exposed to arc flash, arc blast, and shock hazards. He was not wearing rated gloves, because the line was deenergized.

Ken was transported to the local hospital, where he lived for the next few days.

Ken's training was *almost* complete. He remembered all of it except one small point: Voltmeters have limitations.

 ## Test Your Thinking

1. Although records of injuries often include only minimal information, to help prevent similar injuries, the records should include the following:
   a. The type of injury
   b. The type of hazard causing the injury
   c. Both (a) and (b)
   d. Neither (a) nor (b), because the more information recorded, the easier it is for an injured party to sue.
2. A person's judgment generally is based on the following:
   a. Skill, knowledge, and experience
   b. Skill, knowledge, and experience, plus the impact of individual thought patterns

3. The fireball in an arc flash is
   a. Ionized gas (plasma)
   b. A conductive material
   c. Both (a) and (b)
   d. Neither, because the fireball is not a plasma
4. Most incidents and injuries are initiated by
   a. Equipment failures
   b. People
   c. Incomplete procedures
   d. Inadequate regulation enforcement

Notes

1. Ray A. Jones, et al., "Staged Tests Increase Awareness of Arc-Flash Hazards in Electrical Equipment."
2. Mary Capelli-Schellpfeffer, Ralph C. Lee, Mehmet Tomer, and Kenneth R. Diller, "Correlation between Electrical Accident Parameters and Sustained Injury."
3. This account is based on an actual incident. The names, including the name of the facility, have all been changed to protect those involved. Any similarity to actual names or facilities is strictly coincidental.

■ Chapter 6

Protecting the Person

Personal Protective Equipment

Personal protective equipment (PPE) is the final barrier between a person and an electrical hazard. That final barrier is there only if the person chooses appropriate equipment and uses it or wears it. PPE is commonly thought to be protective clothing or apparel—safety glasses and shoes, insulated gloves, or Nomex® clothing. However, PPE also means protective *equipment,* such as rubber blankets, ground-fault circuit interrupters, voltmeters, and safety labels—any equipment that places an obstacle or insulation between the person and the electrical hazard. (See Figure 6–1.)

A person must recognize and understand the hazard before he or she can make the appropriate PPE selection. If exposure to the hazard cannot be eliminated, appropriate protection must be worn. If the hazard is shock, then adequate shock protection must be worn. For example, if the hazard is arc flash, then arc-flash protection must be worn. No personal protective equipment is currently available for protection from arc blast or from flying parts and pieces resulting from arc blast, although some PPE may provide minimal protection.

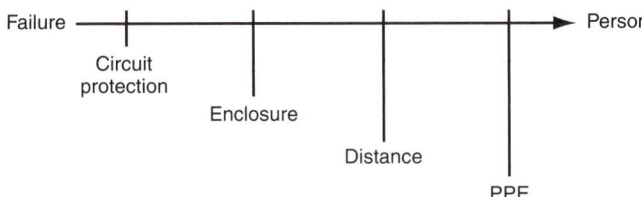

Figure 6–1. Protection Modes.

The type of PPE selected and provided must be based on the hazard as well as both the type and the degree of hazard exposure. Most hazard exposure occurs when operating conditions are *not* normal. However, some exposure does exist when conditions *appear* to be normal. For example, an extension cord can be partially cut and an energized conductor inside can be exposed (or almost exposed). A person might touch this exposed conductor without recognizing that the energized conductor could be touched—without realizing that a hazard exists. It seems that only the way the equipment is installed can effectively deal with this potential exposure.

Standards on Protective Equipment

Tables 6–1 and 6–2 (on page 78) identify consensus standards that provide recommended specifications for protective equipment. Table 6–1, which is similar to Table 3–3.8 in NFPA 70E, *Standard for Electrical Safety Requirements for Employee Workplaces*, 2000 edition, includes references to items generally called electrical protective equipment.

Table 6–2, which is similar to Table 3–4.11 in NFPA 70E, contains references to protective equipment not normally called personal protection.

Rubber products such as gloves, blankets, and sleeves provide protection by adding electrical insulation between the person (an electrical conductor) and an energized point. Because the insulating qualities of these products are extremely important, they must be tested frequently to ensure that necessary test values remain over time. In 29 *CFR* 1910.137, OSHA accepts ASTM standards as the definition of acceptable testing criteria. The ASTM standards are widely accepted for this purpose. However, as with all consensus standards, it is extremely important for the user to identify and consider any limitation of the standard.

Current ASTM standards for gloves[1,2] are intended only for protection from shock. However, experience has shown that the leather protectors also provide adequate protection from thermal energy. In fact, experience shows that heavy-duty gloves with long gauntlets provide adequate thermal protection. Leather gloves normally are stitched together with cotton thread. Experience suggests that the cotton thread will burn away during the exposure but will hold together longer than the arc-flash event.

Table 6–1
Standards on Protective Equipment

Subject	Number and Title
Head Protection	ANSI Z 89.1, *Requirements for Protective Headwear for Industrial Workers,* 1997
Eye and Face Protection	ANSI Z 87.1, *Practice for Occupational and Educational Eye and Face Protection,* 1989
Gloves	ASTM D 120, *Standard Specification for Rubber Insulating Gloves,* 1995
Sleeves	ASTM D 1051, *Standard Specification for Rubber Insulating Sleeves,* 1995
Gloves and Sleeves	ASTM F 496, *Standard Specification for In-Service Care of Insulating Gloves and Sleeves,* 1997
Leather Protectors	ASTM F 696, *Standard Specification for Leather Protectors for Rubber Insulating Gloves and Mittens,* 1997
Footwear	ASTM F 1117, *Standard Specification for Dielectric Overshoe Footwear,* 1993
	ASTM Z 41, *Standard for Personnel Protection, Protective Footwear,* 1991
Visual Inspection of Protective Rubber Products	ASTM F 1236, *Standard Guide for Visual Inspection of Electrical Protective Rubber Products,* 1996
Apparel	ASTM F 1506, *Standard Specification for Protective Wearing Apparel for Use by Electrical Workers When Exposed to Momentary Electric Arc and Related Thermal Hazards,* 1998

ANSI—American National Standards Institute
ASTM—American Society for Testing and Materials
Source: NFPA 70E, *Standard for Electrical Safety Requirements for Employee Workplaces.* Quincy, MA: National Fire Protection Association, 2000.

Types of Equipment

Rubber Products

Until an electrically safe work condition exists, electrical shock protection should be worn that is commensurate with the voltage level of the exposure. In 29 *CFR* 1910.137, OSHA identifies some required electrical rubber products. (Note that the only required electrical personal protective equipment identified by OSHA is listed in this section. Other sections of the OSHA rules prohibit some clothing but establish no other PPE requirements.)

Table 6–2
Standards on Other Protective Equipment

Subject	Number and Title
Safety Signs and Tags	ANSI Z 535, *Series of Standards for Safety Signs and Tags,* *1998*
Blankets	ASTM D 1048, *Standard Specification for Rubber Insulating Blankets,* 1998
Covers	ASTM D 1049, *Standard Specifiction for Rubber Covers,* 1998
Line Hoses	ASTM D 1050, *Standard Specification for Rubber Insulating Line Hoses,* 1990
Line Hoses and Covers	ASTM F 478, *Standard Specification for In-Service Care of Insulating Line Hoses and Covers,* 1992
Blankets	ASTM F 479, *Standard Specification for In-Service Care of Insulating Blankets,* 1995
Fiberglass Tools/Ladders	ASTM F 711, *Standard Specification for Fiberglass-Reinforced Plastic (FRP) Rod and Tube Used; in Line Tools,* 1989 (R1997)
Plastic Guards	ASTM F 712, *Standard Test Methods for Electrically Insulating Plastic Guard Equipment for Protection of Workers,* 1988 (R1995)
Temporary Grounding	ASTM F 855, *Standard Specification for Temporary Protective Grounds to Be Used on De-energized Electric Power Lines and Equipment,* 1997
Insulated Hand Tools	ASTM F 1505, *Standard Specification for Insulated and Insulating Hand Tools,* 1994

ANSI—American National Standards Institute
ASTM—American Society for Testing and Materials
Source: NFPA 70E, *Standard for Electrical Safety Requirements for Employee Workplaces.* Quincy, MA: National Fire Protection Association, 2000.

Because shock has been recognized as an electrical hazard for many years, many consensus standards exist that can help to specify and maintain adequate insulating qualities. The rubber-products industry is well established and performs very well.

Head Protection

ANSI Z 89.1[3] defines criteria for equipment designed and tested to protect a person's head from certain hazards. The standard recognizes and suggests criteria for protection from bump hazards. It also defines criteria for protection from electrical shock as an option. Of course, all head protection worn in the vicinity of energized electrical conductors or circuit parts should have sufficient electrical insulating qualities.

One hazard not addressed by ANSI Z 89.1 is arc flash. In most instances, electrical hard hats are plastic. Although they offer reasonably good protection from electrical shock, the hats offer little, if any, protection from arc flash. The plastic material could easily melt (or ignite) on the wearer's head. When worn within the arc-flash boundary, hard hats and bump caps should be covered by a switchman's hood.

No consensus standard covers construction and testing of equipment that affords arc-flash protection for the head. Protection from arc flash for the head can be had only by using a switchman's hood. The switchman's hood should be selected based on the degree of hazard exposure.

Eye and Face Protection

Standard ANSI Z 87.1[4] covers eye and face protection, but, as with all consensus standards, the scope or boundary of the standard is important. Injury to eyes and faces has been recognized as a danger for many years because projectiles can easily injure these exposed areas. Equipment constructed to meet the requirements of this standard, then, is intended to provide protection from projectile injury.

However, this standard does not address arc-flash hazard. Equipment that is stamped or marked as meeting ANSI Z 87.1 *does not* provide necessary protection from thermal energy associated with this hazard. Protective equipment that *does* provide necessary thermal protection also likely provides protection from impact; however, the reverse is not true.

In an arc flash, released energy comprises frequencies all across the spectrum: visible, infrared, and ultraviolet energy. The objective of eye and face protection is to minimize injury possibilities. In the case of arc flash, several possible hazards are associated with the current flow. The first and most important potential hazard is the thermal energy, which is transmitted by the same mechanics that occur when the warmth from a fireplace reaches a person reclining nearby. The second potential hazard is associated with the energy released in the form of light—visible, infrared, and ultraviolet.

Eye and face protection, then, must protect from these sources of energy. Thermal energy relies on convection, radiation, and conduction. Convection is the movement of heat energy by the motion of a heated material, such as air. Radiation is the movement of heat energy by the direct transmission of energy from a source to a receiver. Air is not required for this type of heat transfer. Conduction is the movement of heat energy through a medium by one molecule being in contact with another; therefore, conduction transfers energy from one molecule to an adjacent molecule.

In the event of an arc-flash incident, most of the energy is released in the form of thermal energy. At least anecdotal information suggests this to be true, based on both the injuries sustained and the equipment destroyed. No experimental data are yet available to either verify or criticize the anecdotal information. It also appears that the majority of the released energy movement is in the form of convection. The speed of an arc-flash event suggests that neither conduction nor radiation has a realistic chance to become an energy transfer mechanism.

Any eye and face protection must be capable of withstanding exposure to thermal energy released in an arc flash. The protective equipment need not necessarily survive the exposure, but it does have to stay together long enough to afford the protection.

Portions of the radiated energy are ultraviolet and infrared. A significant portion of the ultraviolet energy is absorbed or reflected in the face shield and eyeglasses. A larger portion of the infrared energy is transmitted through ordinary face-protection devices. Reflective lenses reduce transmitted infrared energy, but they also reduce the amount of visible light that is transmitted.

Any reduction in the amount of visible light increases the potential of contact with energized conductors, which results in increasing the chance of initiating an arc-flash event. The electrical safety program, then, must weigh the known increased exposure of initiating an event against any potential injury that might result from infrared energy transmitted through the protective material. Perhaps because records may be deficient, the authors can identify no injury resulting from infrared energy transmitted through any face-protective equipment.

Testing of face shields and viewing windows for switching hoods is not yet complete. Some new, highly promising products are being developed. Viewing windows and switching hoods that have an "incident energy" rating are currently being explored in the safety products manufacturing community. For instance, the Oberon Company in New Bedford, Massachusetts, is investigating a material with characteristics that transmit visible light yet will still absorb infrared light.

The authors believe that the critical issue for protection against arc flash is that eye and face protection be worn at all times when a person's head is within the arc-flash boundary.

Footwear

Of course, safety shoes should be worn at all times when a person is on any industrial site, regardless of the existence of an electrical hazard. However, when potential for exposure to electrical shock exists, the role of footwear takes on another meaning. Only in rare instances is the potential hazard other than shock. The idea, then, is to increase the amount of resistance, or impedance. If the footwear is to offer personal protection, then it is important to verify that the protective characteristics are there when needed. A test is necessary. Shoes that are worn as everyday footwear stand a good chance of having their insulating characteristics reduced through the rigors of everyday wear. These shoes should not be worn as electrical shock protection.

The ASTM standard for dielectric boots[5] includes a protocol for testing the insulating characteristics. These overshoes offer protection, provided they are tested and stamped in accordance with the standard. Still, these products should only be used as a secondary isolating method. Appropriately rated gloves and blankets are less subject to degradation through use and should be used as the primary insulating equipment. The best method of protection against hazard, of course, is to establish an electrically safe work condition.

Footwear is an issue from still another point of view. Electronic equipment is susceptible to damage from a small static discharge. To avoid this problem, shoes and other types of conducting devices are readily available for purchase. The objective of this equipment is to provide a conductive path in order to minimize the probability of static buildup on a person who, in turn, could touch the sensitive electronic equipment, resulting in destruction of the electronic circuitry. Footwear or any other conductive device must not be worn where possible contact with open energized conductors exists.

Apparel

Apparel that is commensurate with the type of hazard and the degree of exposure to the hazard should be selected and worn (see Chapter 2, "Electrical Hazards"). A person must execute a hazard analysis in order to develop an understanding of the degree of exposure to each type of hazard (see Chapter 3, "Hazard Analysis"). Exposure to a chemical hazard should result in selection of apparel that affords protection from the chemical hazard. Exposure to a shock hazard should suggest selection of apparel that affords protection from the shock hazard. Selection of apparel for shock protection (e.g., rubber products) has already been discussed.

Apparel that will protect a person from the thermal energy released in an arc-flash incident is a little more complex. As suggested in Chapter 2, thermal energy released in this type of incident increases very rapidly. The longer an arc lasts, the higher the arc temperature becomes. The higher the arc temperature and the greater the degree of exposure, the more the chance of injury increases from the energy release. Apparel must be selected that will protect the person from the degree of exposure.

As the energy level of the source increases, the protective thermal characteristics of the apparel must improve to escape injury. A wide variety of flame-resistant clothing is available. Again, the key concept in selecting apparel is to understand what the apparel will *not* do.

As understanding of the arc-flash hazard grows, professional societies, such as the Petroleum and Chemical Industry Committee (PCIC) of the Industry Applications Society (IAS) of the IEEE, publish papers that document the state of knowledge. Company employees should monitor these and other professional communications and modify their programs as necessary to provide state-of-the art protection for people.

The most significant arc-flash injuries occur as a result of clothing igniting or melting into the skin of the injured person. Some artificial fabrics—generally nylon, polyester, and similar materials—will melt before they ignite. Clothing made from these fabrics should never be worn as external garments. Natural fibers, such as cotton, wool, and silk, may ignite and burn if the material is heated to or beyond the ignition temperature. Melting and/or burning apparel is difficult to remove and subjects the wearer's skin tissue to high temperatures for an extended period of time.

In the vast majority of cases, an arc-flash event is over in less than one-half of one second. Burning clothing extends the event to several minutes. Although the temperature of the burning clothing is much lower than the temperature of the arc, the relative extended time of exposure results in much greater tissue destruction. The duration of the exposure is one component of potential exposure that can be managed by selecting appropriate apparel.

Clothing containing a percentage of non-flame-resistant materials should be avoided, if possible. Underwear normally contains at least some meltable fabrics. Although this type of apparel should be avoided, if sufficient flame-resistant clothing is worn over the meltable material, the temperature of the fabric can be controlled to less than the critical temperature. Underwear made from cotton or another natural fiber remains the best choice.

Table 6–3 offers an illustration of tissue damage when the skin surface reaches the temperature indicated. The idea, then, is to limit the temperature of the skin surface to below these limits.

The ignition temperature of clothing made from natural fibers varies from about 700°F to about 1,400°F. Assuming that the temperature of burning clothing does not decrease while it burns and that one or two minutes are required to remove the burning clothing, the skin surface will surely be destroyed.

Flame-resistant apparel that meets the requirements of ASTM F 1506[6] will not continue to burn after the flame source is removed. Of course, overcurrent devices are intended to remove the source of the flame in an actual arc-flash event. The protective clothing must "hang together" and not be destroyed by pressure forces generated by the arc. The clothing must provide an effective thermal insulation barrier between either the skin or any clothing worn under the flame-resistant PPE and the source of the arc. Several man-made materials, such as the DuPont Company's Nomex®, are available from safety equipment supply houses. It is critical that the protective apparel be purchased to meet the requirements of ASTM F 1506.

Table 6–3
Effect of Temperature on Human Skin

Skin Temperature	Time to Reach Temperature	Damage Caused
110°F	6.0 hours	Cell breakdown begins
158°F	1.0 second	Total cell destruction
176°F	0.1 second	Second-degree burn
200°F	0.1 second	Third-degree burn

Voltmeters

In most cases, a voltmeter is considered to be a tool in the same vein as a screwdriver. Although many people carefully consider the quality of their hand tools, voltmeters are surely in a different class of equipment. Frequently, a voltmeter is the device that provides sufficient warning that a conductor is energized, averting potential disaster. The voltmeter should be considered as PPE and used as such. Only the highest quality devices should be purchased. After purchase, adequate storage and maintenance should maintain the meter as close to as-purchased condition as possible. (See Figure 6–2.)

Consensus standards published by both IEEE and UL® contain important requirements that aid in avoiding voltmeter-related incidents. Many incidents occur because a voltmeter is set on the incorrect scale. Many others occur when the leads slip from their plugs on the meter surface. Still others occur when an internal component failure initiates a large arcing fault inside the device. In some instances, voltmeter leads have a voltage rating less than the maximum scale voltage. Most of these incidents could be eliminated simply by using only those meters that meet IEEE and UL requirements.

Some solenoid-type voltmeters are designed for and assigned a "duty cycle." The duty cycle means that a time limit must not be exceeded. In most cases, the duty cycle is identified on the label, and, in other instances, that information is available only in the catalog information. If the duty cycle is exceeded, the device can literally explode in the user's hand.

The real issue seems to be that a person using a voltmeter *must* understand any limitations or boundaries associated with the device. The

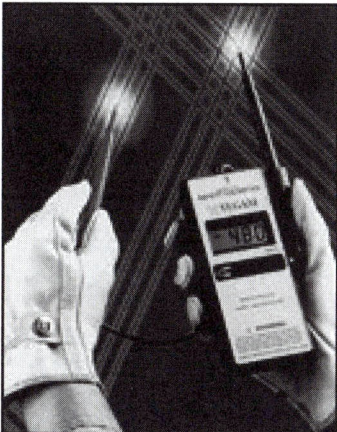

Figure 6–2. Tegam 110 Amps (Voltmeter).

user must be knowledgeable about how to use the device and how to interpret meter indications. Training in the use of the chosen voltmeter is imperative.

Labels and Labeling

Manufacturers normally install labels that provide information related to application of their equipment. Much of the information provided on the label is required by some consensus standard. The *National Electrical Code®* (*NEC®*) requires some label information, NEMA standards require other information, and third-party testing laboratories require still other information. However, all required information is related to the application of the equipment. Information related to how people interact with the equipment or service generally is missing from the label. (See Figure 6–3.)

Information that might help people interact with the equipment or service should be readily available at the equipment location. For example, equipment should be labeled to indicate the approach boundaries. If protective equipment is required to operate the disconnecting means, a sign or tag should warn anyone approaching the equipment that PPE is required. The intent of the added labels or tags should be to ensure that sufficient warning is offered to people as they approach.

WARNING

ARC-FLASH PPE REQUIRED

Available incident energy is 19 CAL/CM ^{2}

Figure 6–3. Equipment Information Label.

Ground-Fault Circuit Interrupters

Several different references suggest that a person's body can sustain an electrical shock of 6 mA without injury. Therefore, if a shock could be limited to that amount of current, a person could be protected from injury. In the mid-1970s, a device called a ground-fault circuit interrupter (GFCI) was manufactured and marketed. In the early days, a GFCI was susceptible to false trips and was generally regarded as inaccurate. Some criticism was justified at that time. Manufacturers were responsive to the electrical community and solved the design and manufacturing problems.

GFCIs work by detecting ground faults and immediately switching off the power to the circuit. They compare the amount of current flowing in the "hot" conductor with the amount in the neutral, "grounded" conductor. Should an imbalance between these two current measurements occur that is greater than 6 mA, the GFCI removes the source of energy in a short period of time. The GFCI assumes that an imbalance is caused by some of the current supplied to an item of equipment returning through a person's body, because the current is not detected in the neutral conductor.

The GFCI constantly monitors the current flowing in the circuit. If the current going into the circuit differs even slightly from the current returning (leakage), the GFCI immediately switches off power to the circuit. A GFCI detects very small variations in the amount of leakage current, even amounts too small to activate a fuse or circuit breaker. GFCIs work quickly enough to protect against serious injury and electrocution.

Several different types of GFCIs are currently in use:

- *A circuit-breaker GFCI* can be added to an electrical panel to replace ordinary circuit breakers and protect selected circuits. A circuit-breaker GFCI shuts off the electricity in the event of a ground fault and also trips when a short circuit or an overload occurs [see Figure 6–4 (A)].
- *A wall-receptacle GFCI* can be used in place of a standard receptacle found throughout a house that is protected by either fuses or circuit breakers. It fits into a standard outlet box and protects against ground faults whenever an electrical product is plugged into the outlet [see Figure 6–4 (B)].
- *Plug-in* and *portable GFCIs* simply plug into any receptacle and require no special knowledge or equipment to install. One type features the GFCI circuitry in a self-contained enclosure with plug blades in the back and receptacle slots in the front. It can be plugged into a receptacle and then an electrical product plugged into the GFCI. Another type of portable GFCI is an extension cord combined with a GFCI. Such a GFCI adds flexibility in using receptacles that are not protected by GFCIs [see Figure 6–4 (C and D)].

The *NEC* identifies some locations where a GFCI should be installed. Other standards suggest that a GFCI should be used in any area where water or moisture exists. GFCIs are inexpensive. They are available in several different constructions: receptacle type, portable type, or circuit-breaker type. Each construction is effective and can prevent injury if properly used. The 6-mA current might be perceptible to some people. The current might be in excess of 6 mA. Although it is possible that the person might be shocked, the body reaction will be limited to surprise. No electrocution will occur. (See electrical shock discussion in Chapter 2, "Electrical Hazards.")

GFCIs are designed and tested to fulfil requirements of UL 943.[7] This standard requires the voltage source to be removed when the current imbalance is measured to be between 4 and 6 mA. The GFCI is also required to contain a "push to test" button that verifies the mechanical operation of the GFCI. It was suggested earlier that current flow through a body is unlikely to remain at the same level. Current

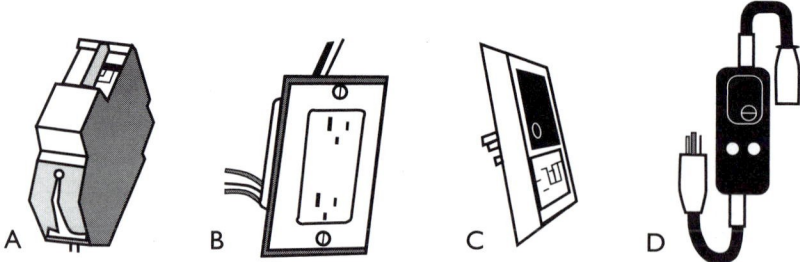

Figure 6–4. Types of Ground-Fault Circuit Interrupters.

flow tends to increase with duration of contact. Because a GFCI is a mechanical device, time is required for it to operate. For a GFCI to prevent electrocution, the speed with which it operates is important. The operating speed varies, indirectly, with the amount of current imbalance. When properly used, a GFCI can certainly prevent injury.

GFCIs are no longer exotic and expensive protective devices. No longer are they subject to nuisance tripping. Instead, they are very reliable and cost effective and should be installed where receptacles are frequently used to supply tools and equipment.

Overcurrent Devices

Overcurrent protection is intended to remove the source of energy from a circuit in the event of a short circuit, an arcing fault, or other failure. The *NEC* includes many requirements for selecting and sizing overcurrent devices. It is important to remember that the *NEC* is a *minimum* standard. In other words, requirements contained in the *NEC*, in reality, are negotiated to achieve a consensus. Each requirement is negotiated among the talented and knowledgeable panel members. There is certainly nothing wrong with an installation that exceeds these consensus requirements.

In the case of personal safety, the amount of time required to sense and remove the energy source is critical to minimizing any potential arc flash, pressure wave, or equipment destruction. Faster operation of overcurrent devices means less energy is released in the fault. If the released energy is reduced, then potential injury is commensurately reduced. In fact, energy is related to the square of the time.

In the past, overcurrent protection has been selected and installed with the intent of managing equipment damage. In fact, overcurrent protection should be selected and installed on the basis of minimizing any potential injury to people. If potential damage to people serves as the basis for selecting overcurrent protection, damage to equipment will be at a minimum.

In general, overcurrent devices that sense a problem and then initiate a mechanical action are much slower than devices that inherently sense and act simultaneously. Circuit breakers generally require some mechanical motion to remove the energy. In many instances, 6 cycles (and sometimes 30 cycles) are required to remove the source of energy. Fuses typically employ the use of a melting alloy-type of material. When the alloy melts, the circuit is opened. Depending on the type of fuse, the arcing time can be a few cycles. Current-limiting fuses limit the arcing time to less than 2 cycles, provided the fault current is within the current-limiting range. Of course, the actual time needed to clear a fault depends on the degree of the fault and can be determined by consulting the manufacturer's time-current curves.

 # A Closer Look

Hans said goodbye to his wife and four-year-old daughter as he left for his job at the textile plant in Dusseldorf.[8] He had worked at the plant since completing the prescribed course at the local trade school three years earlier. After graduating from the trade school near the top of his class, Hans received a salary increase. He was a licensed electrician.

In his three years as an electrician at the plant, Hans' dedication and work ethic had earned him complete respect from the plant management as well as from his coworkers. Hans paid close attention to all procedures and work practices at the plant.

This particular Thursday morning, Hans planned to perform annual maintenance on a substation secondary circuit breaker. The maintenance was at least 12 months overdue, a fact that had concerned Hans. After checking with production, he proceeded to substation 2-3, which contained the circuit breaker that needed maintenance. Hans observed all the instrumentation and saw that everything seemed normal. He stopped moving and listened carefully for sounds that seem unusual. There were no unusual sounds, smells, or other indication of a problem.

The unit substation contained two transformers but only one set of loads. The transformers were in phase with one another. This arrangement permitted the loads to be fed from either transformer to keep the plant running. It was common practice for a load to be transferred from one transformer to the other. The transformers were paralleled for a short period of time—about two seconds—which was long enough for a person to close one breaker and then open another. The transformer secondary was a solidly grounded wye configuration. The only existing overcurrent protection for either transformer was in the primary. The secondary voltage in this plant is 380 volts ac.

The plant had revised its procedures to require arc-flash protection at any time a compartment door was opened until the compartment was tested "dead" (an interesting choice of words). Hans knew what the procedure required. In fact, he wore his 6-ounce Nomex® coveralls all the time. The procedure required a polycarbonate face shield. Thinking that he was fulfilling the requirements of the procedure, Hans reached for the nearby face shield and went back to his work. Hans' face shield was, in fact, ordinary clear polyurethane. The procedure also required a second person to be nearby until the compartment was tested "dead." Hans asked another electrician working in the same room to stand by, and he agreed.

Hans closed the breaker in the top compartment. The two transformers were now in parallel. Hans was always a little nervous while the transformers were in parallel. He knew that the substation bracing was not rated for the capacity of both transformers.

Next, he opened the circuit breaker in the bottom compartment and immediately tested for voltage. His intent was to disconnect the circuit breaker from the bus and remove it for testing and cleaning. Disconnecting the circuit breaker meant inserting and operating a hand crank with the compartment door wide open. Hans put on his gloves and began to crank.

Just as the circuit breaker started to move, a major arc flash occurred. The movement inside the compartment had initiated a fault. Hans was completely encircled in the resulting ball of fire. The overcurrent protection operated even better than expected. The fault was removed in 7 cycles. The standby person did not see the ball of fire. He could not turn his head fast enough. He did hear the noise, however. In fact, the noise rang in his ears for several hours.

The polyurethane face shield that Hans was wearing turned black and rolled up around the top of his head. The cotton stitching in his leather gloves burned away, and the leather fell to the compartment floor. The chest area of Hans' Nomex® coveralls was black and brittle. The coveralls were also blackened on the front of one leg.

Hans had a burn on the front of his right thigh where the Nomex was tight against his skin. His cotton undershirt was undamaged. His face looked as if it were sunburned. The face shield could easily have ignited, but it didn't. The gloves prevented any hand injury.

Even with the wrong equipment, Hans had escaped. He was very glad to see his wife and daughter that afternoon.

Test Your Thinking

True   False

☐   ☐   1. Although designed for shock protection, experience has shown that voltage-rated gloves with leather protectors provide protection from thermal energy.

☑   ☐

☐   ☐   2. Thermal energy relies on convection, radiation, and conduction to transfer energy.

☐   ☐   3. As the energy level of the source decreases, the protective thermal characteristics of the apparel must improve to escape energy.

☐   ☐   4. Faster operation of overcurrent devices means more energy is released in the fault.

Notes

1. ASTM F 496, *Standard Specification for In-Service Care of Insulating Gloves and Sleeves*, 1997.
2. ASTM F 696, *Standard Specification for Leather Protectors for Rubber Insulating Gloves and Mittens*, 1997.

Answers: 1. (true), 2. (true), 3. (false), 4. (false)

3. ANSI Z 89.1, *Requirements for Protective Headwear for Industrial Workers*, 1997.
4. ANSI Z 87.1, *Practice for Occupational and Educational Eye and Face Protection*, 1989.
5. ASTM F 1117, *Standard Specification for Dielectric Overshoe Footwear*, 1993.
6. ASTM F 1506, *Standard Specification for Protective Wearing Apparel for Use by Electrical Workers When Exposed to Momentary Electric Arc and Related Thermal Hazards*, 1998.
7. UL 943, *Ground-Fault Circuit Interrupters*, 1993.
8. This account is based on an actual incident. The names, including the name of the facility, have all been changed to protect those involved. Any similarity to actual names or facilities is strictly coincidental.

■ Chapter 7

Standards

Types of Standards and Their Objectives

One understanding of the term *objective* is that a person is trying to reach a goal. In the case of an organization, the term *mission* defines the goal. All standards published by a standards-developing organization, then, should be read in light of the organization's mission.

In some instances, the standards-developing arm is a subset of a larger organization. For example, the Institute of Electrical and Electronics Engineers (IEEE) is a worldwide institution that focuses on electrical engineering interests. Those interests include objectives such as continuing education and generating new knowledge. However, one major objective of the IEEE is to achieve standardization among manufacturers and industries and across other boundaries.

Based on that understanding, a standard cannot have an objective. Instead, a standard is a tool that is used by a standard-developing organization (SDO) to help achieve its mission.

A standard can also be roughly defined as a product that has the same or similar physical characteristics. This meaning enhances the ability of one manufactured product to be exchanged for another. A standard can also be defined as a *normal* method of doing something. However, other terms, such as *procedure* and *practice,* may also have similar or identical meanings. Without further definition, little meaning can be derived from the term *standard,* since so many different connotations may apply.

Standards have a scope, which defines and sometimes discusses the primary objective of the standard. However, perhaps the most important aspect of an adequate scope is to define the limits of the document. Frequently, a person reads a scope within a standard and does not understand that he or she is also reading about boundaries. Standards sometimes define the boundary or limit of the standard's intended application by including exception statements. For example, Article 90-2 of the *National Electrical Code®* (*NEC*) has an exclusionary statement covering installations that are under the exclusive control of an electric utility. But a similar exclusionary statement in OSHA 29 *CFR* 1910.304 has a different twist. A good understanding of the scope is necessary in order to apply any standard as intended by the SDO.

OSHA Standards—Codes of Federal (U.S.) Regulations

The U.S. Occupational Safety and Health Administration (OSHA) is chartered by the legislative process as Public Law 91-596. Although the law has been modified a couple of times, it remains essentially the same as initially considered to be the Williams-Steiger Act. Modifications to The Act have related to sharpening the focus of the affected communities. For example, a modification affected by congressional action in November 1990 embraced memorandums of understanding (MOU) between the U.S. Department of Labor (OSHA) and the U.S. Department of Energy. Until the MOU became effective, the Department of Energy facilities were not guided by OSHA regulations.

Congress declared that the purpose of The Act was to ensure safe and healthful working conditions for every working woman and man in the United States, as far as possible.[1] In addition to defining the purpose of Section 2, The Act defines *how* OSHA must achieve the objective. The Act also contains language indicating an expectation that employers and employees need to work together to reduce the number of hazards. It suggests that responsibilities of employers and employees are clearly separate, but interdependent.

Section 2 authorizes the U.S. Secretary of Labor to do the following:

- Set standards applicable to businesses affecting interstate commerce.
- Create an Occupational Safety and Health Review Commission for adjudicating disputes related to The Act.[1]
- Develop innovative ways to deal with safety and health problems, including research.
- Define a medical criterion that "practically" avoids diminished capacity or life expectancy.

- Develop and promulgate "standards," and OSHA is strongly encouraged to use "consensus standards" where they exist.
- Provide an effective enforcement program.
- Encourage states to assume responsibility for their own safety and health programs.
- Establish records, reports, and reporting procedures, as necessary, to describe the nature of the occupational safety and health problem.

The Act requires the Secretary of Labor to encourage joint labor management efforts. In fact, experience supports this intent. Safety programs that joint labor and management encourage generally are effective.

The OSHA technical staff produces "standards" that are promulgated in the *Federal Register*, as it is the government's method of communication with the public. Section 6 of The Act defines the process whereby standards produced by the technical staff become enforceable as rules.

From the initial organization, OSHA attempts to align its standards with segments of industry or trades. Representatives of industry or trade argue through the 6(b) process (of The Act) that their particular segment should be exempted or have the standard modified. For example, electrical hazards are addressed in many different places in the OSHA standards. In response to lobbying efforts, OSHA attempts to address legitimate arguments and provides some segments with its own set of rules.

The basic physics of how people are injured does not change from one industry, discipline, or trade to another. An injury is related to the type of task being performed. It is not related to the trade or discipline of the employee. For example, falling from an elevated platform will likely result in an injury. An electrician or a machine operator can be electrocuted by defective ground-fault protection. All injuries result from exposure to hazards. If hazards differ among trades or disciplines, the difference is usually in terms of the degree of exposure or how the person is exposed.

The National Electrical Code®

Instead of a set of standards, as in OSHA, *NEC* is a single standard. Technical committee members from industry volunteer their time and efforts to write this standard. The publisher is the National Fire Protection Association (NFPA). Because the *NEC* is a standard, it has a scope rather than an objective. Like OSHA, the NFPA has an objective. Also like OSHA, NFPA standards are tools used to accomplish the objective or mission of the organization.

The NFPA was organized in 1896 to improve the quality of life by reducing fires. The association assumed responsibility for a standard covering the installation of sprinkler systems. The mission grew and changed over the years as needs for standardization changed. The current mission of the NFPA includes promoting the science of and improving the methods of fire prevention and protection, electrical safety, and other safety-related goals. The initial intent to improve the quality of life remains today.

The *NEC* is a document that covers installation. The essential focus of the *NEC* is to define how equipment should be installed. Eliminating fires by identifying installation requirements remains a large part of the *NEC* content. Any service or equipment installed that meets the requirements identified in the *NEC* will be safe while the equipment is operating normally, provided it is adequately maintained.

As published by the NFPA and the American National Standards Institute (ANSI), the *NEC* is only advisory. The NFPA is a developer and publisher of standards. However, the *NEC* is written in language that can be readily inspected and enforced. The *NEC* becomes mandatory when a governmental body adopts it. The enforcing authority is authorized by, and defined in, the law or ordinance that embraces the *NEC*.

NFPA 70E

While supporting the general objective of NFPA discussed previously, the prime objective of the 2000 edition of NFPA 70E, *Standard for Electrical Safety Requirements for Employee Workplaces*, 2000 edition, seems to be different from the 1987 edition. The initial history of NFPA 70E is closely associated with a request by OSHA to produce a standard covering work practices. Therefore, the prime objective of the standard is to provide electrical safety-related work practices.

The objective of NFPA 70E is very similar to the objective of OSHA 29 *CFR* 1910, Subpart S, with the exception that NFPA 70E is a *consensus* standard, and Subpart S is not. NFPA 70E is written and published through the same consensus process as each of the other NFPA standards. That process ensures equal participation by all interested sectors of the community. Adoption as an American National Standard certifies that the process is open and accessible.

Of course, change or review of either a new or revised OSHA standard, such as 1910, Subpart S, must follow the rules detailed in Section 6 of the OSH Act. On the other hand, the NFPA process for review and change is much more accessible to the public. NFPA 70E can be reviewed and revised as necessary to retain its relevancy. The result of the ease of revision is that NFPA 70E addresses all hazards and work practices that are currently generally accepted by the community. For example, in Subpart S, OSHA contains neither information nor requirements for protection from either arc flash or arc blast. On the other hand, NFPA 70E does address these hazards and offers suggestions about how to afford protection for people. NFPA 70E is considerably more protective from electrical hazards than OSHA standards.

The National Electrical Safety Code

The IEEE publishes the *National Electrical Safety Code* (*NESC*). Although substantially different from the NFPA standards process, the IEEE process is accredited by ANSI. The process ensures that affected interests are represented.

The objective of the *NESC* is similar to that of the *NEC*. However, the scope of the document is quite different. The *NESC* addresses the utility industry and provides information intended to protect the general public from hazards associated with the transmission and distribution of electricity.

Transmission lines provide services to large segments of the public, including medical and other emergency facilities. Maintaining electrical service for emergency facilities is, of course, very important. Industrial facilities frequently require uninterrupted electrical service by contract. Work rules, as contained in the *NESC*, are intended to afford protection for utility workers as they go about their tasks of maintaining systems while the lines are energized, as is frequently necessary to maintain uninterrupted service. The attempt to provide uninterrupted business service effectively defines the working environment for many utilities.

A working environment that expects and encourages employees to accept exposure to electrical hazards should be avoided. The authors are not suggesting a weakness in the *NESC*. Instead, the process by which an employer adopts and applies the standard is very important. Because the *NESC* defines methods to interact with energized transmission and distribution lines, most employers simply adopt these work methods as a "norm."

Employees should not be expected to accept unnecessary exposure to hazards. The requirements of all consensus standards, when adopted, should be implemented with full knowledge of how people are exposed to hazards. The working environment should consist of processes and work methods that minimize or avoid exposure to hazards.

National Electrical Manufacturers Association Standards

One result of using standards is that it affords a common base for product interchangeability. For example, without a "standard" configuration, each purchase of an electrical tool would necessitate exchanging the cord cap before the tool could be used. Without "standard" ratings, the internal construction would vary from one product to another.

The National Electrical Manufacturers Association (NEMA) is a nonprofit standards-developing organization. Many of NEMA's standards are accepted as ANSI standards. The primary objective of a NEMA standard is to promote interchangeability among products. As in all electrical standards, the objective of NEMA standards is to provide for minimum acceptable safety.

NEMA is an association of manufacturers of electrical products. As such, one intent of the association is to avoid excluding any manufacturer. However, the standard defines requirements that products must meet.

Underwriters Laboratory Standards

Underwriters Laboratory® (UL) is a nonprofit organization whose prime reason for existence is testing the safety of products to protect the general public. Of course, UL tests products from many other disciplines. Before any test can be conducted, a test protocol must be established. That protocol must provide the necessary information to derive a satisfactory conclusion. That test protocol is a standard. In many cases, the UL standard is accepted as an ANSI standard.

A test engineer typically generates a UL standard as a test protocol in response to a request to test a product. Frequently, the test protocol becomes a "consensus standard" by the canvass method. The objective of a UL standard is to determine if a product will perform a function as defined by the protocol or standard. Therefore, knowledge of the testing standard content is important. The idea is to resist simply adopting any UL standard on blind trust. Instead, the user must become acquainted with the standard's scope and objective.

Other consensus standards-developing organizations function similarly to UL. For example, Factory Mutual produces testing standards. Although the organizational objective is different from UL, the standards products have a similar objective.

American National Standards Institute Standards

The American National Standards Institute (ANSI) is not an SDO. ANSI does not develop standards. Instead, ANSI monitors the development of standards by other organizations. ANSI focuses on *how* the standard was developed. If the standard-developing process meets criteria developed and published by ANSI, the standard can be accepted as an American National Standard. Of course, only one American National Standard can be accepted for each subject.

The standards process is voluntary in the United States. Whether a standard has a direct legal status completely depends on whether the standard has been adopted by statute within an established jurisdiction or not. An employer, a governmental body, or other organization may choose to embrace and implement any particular standard.

Like many other organizations involved with the production of standards, ANSI is a voluntary not-for-profit organization. It is not directly associated with government. However, the National Institute of Standards and Technology (NIST) has identified ANSI as the official U.S. representative to international standards efforts.

The ANSI objective, then, is to ensure a complete and open process for organizations that produce standards. ANSI ensures that rules associated with due process have been followed. In other words, every interested party has the opportunity to have his or her opinion considered in the development process.

Standards Outside the United States

A standardization effort exists in virtually all developed countries. The general objective of these efforts is similar to standardization efforts in the United States. The need for standards is the same the world over (i.e., interchange of equipment, safety and health of people, safety of the public, etc.).

In most cases, standards either facilitate trade (by providing for interchange of equipment) or hinder trade (by establishing requirements that products from another country do not meet). International standards attempt to reach consensus among people who represent interests across governmental boundaries. Obviously, these international standards present an opportunity for enhanced commercial trade.

In most instances, standardization efforts are either initiated or embraced by federal governments, whereas in the United States all standardization efforts are voluntary. Many efforts to produce standards are primarily concerned with commerce. Many standards tend to contain some element of commercial protectionism related either to a country, an industry, or a manufacturing consortium.

Naturally, as people write or participate in writing a standard, the content tends to be based on their personal experiences. However, with sufficient technical knowledge and a willingness to focus on *needs* rather than *desires,* a technically effective standard can be produced. This point underscores the need for people to be involved in the process of generating standards that they consider to be important.

Efforts to produce standards that are widely accepted in the United States are essentially mirrored by efforts with similar objectives. For example, third-party evaluation, facility inspection, product standards, installation standards, and maintenance standards exist in the United States and also across developed countries. In general, safety, health, and environmental standards exist in developed countries as well. The objective of these efforts essentially is the same as the objective of safety, health, and environmental standards in the United States. The names, processes, and requirements are different; however, the objective remains the same: prevent injury to people in the workplace.

In the electrical discipline in the United States, the OSH Act attempts to prevent injury by embracing requirements defined in national consensus standards, where they exist. As mentioned earlier in this chapter, OSHA defines work practices and requires employers to implement those practices. Employers are required to ensure that their employees are trained. The strength of this program, then, is in the technically justified procedures that are used in the workplace.

In many other countries, comparable regulations tend to emphasize training requirements for employees. The process generally involves a system of certification for qualified workers, including definitive training mechanics. Employers, then, are required to ensure that workers are appropriately licensed for their trade. The program strength is related to this system of producing qualified persons through training.

Because standards are related to the experience and knowledge of the people who write the standards, significant differences can exist. Equipment integrity and component testing vary from one country to another. Grounding requirements and lockout requirements consider all other technical issues in the experience of the people writing the standard. For example, at some voltages, standards may require disconnecting means to automatically earth conductors in some situations. Design and maintenance processes must consider these variations.

Internal Corporate Standards

In some instances, large corporations establish and operate standardization efforts that apply only to internal organizations. Internal programs usually have two objectives: control costs by defining specific engineering choices, and control exposure to safety, health, and environmental hazards. The documents produced by these efforts are sometimes called standards. They might also be called guidelines or procedures. However, the internal product usually serves in the same capacity as a standard.

Corporate legal organizations are frequently concerned about the name assigned to internal standardization efforts. Their concern is that if documents are called "standards," a disinterested party might expect all requirements of these standards to be implemented in all instances and in all locations. The legal concern is the potential use of the internal standards in litigation.

In the electrical discipline, programs that produce internal standards tend to select appropriate national consensus standards and offer interpretations of those requirements. If the external standard allows for a choice, the internal standard might make the selection for the corporation. Internal programs normally result in a high-quality installation and an excellent maintenance program.

Interaction of Standards

Standards developing organizations typically produce three different types of documents: standards, guidelines, and recommended practices. Although each product document might be called by a different name, the products tend to fall into these general categories. A standard is intended always to be applied and uses words such as *shall* and *must* to define requirements. If words such as *should* or *may* are used in conjunction with an idea, the standard implies that a choice by the user is intended.

When a standard is embraced and implemented, the body or unit that implements the standard also determines how the standard is to be interpreted and enforced. If the standard is implemented by regulation, the same regulation must also identify who is accountable for interpretation, inspection, and enforcement. The term *authority having jurisdiction* is frequently used in standards' language to mean "this accountable agency." It follows, then, that if a standard becomes legally binding, the regulation overrides any owner-enforcing authority. However, if an owner adopts a standard (where no regulation exists), then the owner must also identify the mechanism for inspection and enforcement.

One element of an electrical safety program should identify all standards that are embraced by an owner or employer. Of course, standards that have been adopted by regulation should also be identified in the electrical safety program (see Chapter 10, "The Electrical Safety Program").

Monitoring Federal Regulatory Requirements

In the United States, the federal government is assigned authority by the Constitution to regulate interstate commerce. It is from this platform that the Williams-Steiger Occupational Safety and Health Act (The Act), assigns the charge of establishing and operating the Occupational Safety and Health Administration to the U.S. Department of Labor.

The *Federal Register* is the tool by which the federal government communicates (or, in most cases, attempts to communicate) with the public. The *Federal Register* documents contain the planned and executed actions of the Department of Labor, as well as all other federal actions that affect the general public. Scanning a document is necessary for interested persons to be continually aware of regulatory actions of OSHA or other departments that have a bearing on electrical safety.

The *Federal Register* is the place where the 6(b) process contained in The Act is implemented. It is the communication medium used by OSHA to announce intended and preliminary drafts of standards. Citations of *Codes of the Federal Register* (*CFR*) should always specify title, part, and section number, in that order. Thus, 29 *CFR* 1901.1 refers to Title 29, Part 1901, Section 1.

Because the *Federal Register* is the official communication medium, an owner or employer should scan each edition for indications of new or revised regulations. The *Federal Register* is available for purchase both electronically and by paper mail. The World Wide Web site is http://www.OSHA.gov. Paper copies may be ordered from the U.S. Government Printing Office, 710 N. Capitol Street, NW, Washington, DC 20401. Compact disks of OSHA rules are available for purchase at: www.access.gpo.gov/su_docs/sale/sale300.html. Several commercial organizations scan the *Federal Register* and then market that information.

Monitoring State and Local Regulatory Requirements

Regulatory processes vary widely from one locale to another. Even geopolitical units vary. As a result of these variations, no methodology is universal. The only practical method for an owner or employer is to frequently monitor city and county discussion. Local newspapers usually report the subject matter covered by these units. It is wise for an employer to establish a routine process that provides information for interested individuals about regulations adopted by these local authorities. In many instances, local civic organizations such as user councils provide insight regarding local governmental actions. Frequently, local politicians rely on civic units to build understanding of public needs.

Determining Employer Policies

Every organization exists for a reason. Industrial and commercial organizations generally exist to make money. However, these organizations are made up of people who maintain values and beliefs. Safety policies are usually based on some element of the employer's values and beliefs. In this instance, *employer* values and beliefs refer to the composite management structure.

In many instances, employer's policies are driven solely by legal concerns. Establishing a sound defensive legal posture is important; however, effective safety policies are more important. Some employers believe that personal safety needs override the desire to establish a strong defensive posture. In these cases, the employer does not ignore legal concerns; instead, the company expends the greatest effort and amount of money to prevent incidents and injuries.

If a strong legal defense is determined to be the highest priority of a corporate policy, the standards policy should key on legally mandated regulations. Actions and discussions should reinforce this policy. On the other hand, if preventing injury is considered the highest priority, the policy should identify consensus standards (both legally mandated and not mandated) that the employer considers best able to provide adequate guidance to this end. Again, real implementation of the standards policy involves the entire organization.

It is important to note that legally mandated standards uphold the objective of preventing injury. Governmental units exist only to serve the public. When these units mandate a standard, such as the *NEC*, their intent is public safety. In many instances, the intent of legally mandated standards is to prevent injury to members of the general public. In some instances, such as OSHA, the intent is to prevent injury to industrial employees. The scopes of consensus standards must be reviewed to understand the mission. (For more information, see Chapter 5, "Strategies for Preventing Injury.")

Current editions of the following standards are recommended for inclusion in an employer's electrical standards policy:

- NFPA 70 (*NEC*)
- NFPA 70B (*Recommended Practice for Electrical Equipment Maintenance,* 1998 edition)
- NFPA 70E
- NESC
- OSHA Section 1910.147 of Subpart J
- Subpart S
- Section 1910.269 of Subpart R

Employers with international facilities must consider international standards as they review SDOs for applicable products. Many countries have standards programs that are mandated by governmental action. Employers with international facilities must determine which standards apply to each facility. The standards policy included in the electrical safety program for each site should identify which international standards apply on that site.

Monitoring Activity of Standards-Developing Organizations

The system of consensus standards in the United States is voluntary. Volunteers working within the guidelines of a nonprofit organization write the standards. The SDO then publishes the finished product. The organization of the process requires that a communication mechanism be in place. Communication is necessary within the volunteer community. For the product (standard) to be considered as a national consensus standard, the communication system must be an open process that allows input from the public.

In most instances, the only action necessary for a person to be included in the communication process is for the person to request it. Any person who volunteers or otherwise participates in the process of writing or commenting on a standard receives all necessary information to be continually aware of a standard's state of issue.

In most cases, SDOs maintain sites on the World Wide Web. Web sites are rapidly becoming the primary mechanism for communication between the general public and the SDO. Regularly visiting the Web sites of SDOs is an excellent way to keep abreast of SDO plans and activity.

Generally, all proposals and comments are welcome, and it is not necessary to be a member of the organization to participate in the standards process. However, the most effective participation is to attend committee meetings and other necessary meetings of the organization. Funds for these associated travel expenses sometimes are significant. Time and energy are also required, which normally translate to funds.

In most instances, participation funding is provided by industrial organizations. Sometimes, the organization is an industrial organization; other times, the organization is another type of unit, such as a manufacturer. In a few instances, an individual provides necessary funds, but this instance is rare.

The authors strongly recommend that company representatives participate in the process of writing standards—not all standards, of course—but certainly all standards embraced by their companies' policies. These standards contain the collective best thinking and experience of the committee members. Any person or organization that does not participate in the process is shortchanging the organization.

Scanning SDO Lists for Applicable Products

By scanning the published list of SDO committees, a company representative can easily identify a few potential standards that cover any issue being considered. After identifying particular committees of interest, the representative can scan each standard produced by that committee.

In some instances, several standards might address a single issue from different points of view. In this case, the SDO usually tries to ensure that the content of each standard is correlated with the content of each of the other existing associated standards. For example, in the NFPA process, several standards cover electrical issues:

- NFPA 70 covers electrical installation requirements.
- NFPA 70B covers electrical equipment maintenance.
- NFPA 70E covers electrical safety-related work practices.

It is important that these standards do not contradict one another. To have the best chance of avoiding contradictions, each panel or committee that writes these standards reports to NFPA through the Technical Correlating Committee (TCC). The TCC is charged with reviewing draft products of each panel or committee with an eye toward eliminating any contradiction. The TCC cannot correct any identified problem. Instead, the originating panel or committee must resolve identified problems.

Selecting Interpretation Mechanics

Because recommended practices, guidelines, and standards are written by committee action, the intent of some requirements may be somewhat unclear to a reader. A reader's interpretation tends to be influenced by personal experience. The text of a requirement is the result of negotiation among committee members with opposing opinions. The final text sometimes may be a bit convoluted. The point is that some interpretation of a requirement may be necessary.

In some standards, an authority having jurisdiction (AHJ) is identified as the final authority for interpreting any particular requirement. Without an AHJ, disagreements associated with a requirement would be difficult, or even impossible, to resolve. Generally, the body that implements a standard identifies the AHJ in the regulation or policy. Of course, a standard that is required by a governmental unit identifies the process for enforcement. With enforcement comes the authority for interpretation.

When an employer embraces a standard, the employer then must also identify the enforcement mechanics. Like mandated standards, interpretation is then assigned to the enforcement unit. Each standard must be enforced as a "standard" method. The AHJ also should be identified in the standards policy.

A Closer Look

The Wendstedt Company[2] was considered to be Northern Ireland's safest plant. In its 40 years of existence, the company had consistently posted the lowest injury statistics in the United Kingdom. In fact, Wendstedt's safety program was considered the benchmark and was the envy of many other corporations.

Jack, an electrical foreman, was a second-generation worker at the plant. He had worked at the plant for 28 years. His father, who had retired recently, worked there for over 40 years. Neither Jack nor his father had ever been associated with an incident in which someone received any significant injury. Their combined experience asserted that Jack was very safe while on the job. However, that experience was about to change.

On a typical Wednesday afternoon, the production supervisor on the day shift called Jack to tell him about a problem with an extruder. The extruder was driven by a 100-horsepower, 380-induction motor. This motor was critical to continuous operation of the intermediate product. After talking with the extruder operator, it was clear to Jack that the problem was with the electrically operated circuit breaker being used as a starter.

Because the motor was critical, an alternate supply was installed. The alternate supply simply paralleled the load side of both the primary and alternate circuit breakers. The alternate breaker was permanently installed so that the alternate supply could be operated quickly in order to avoid significant product loss.

The applicable standard called for the conductors supplying the motor to be automatically earthed when the motor was locked out for service or maintenance. However, because the motor had two sources of energy, the load side of the "earthing" switch at the alternate circuit breaker was electrically connected to the earthing switch for the primary circuit breaker.

The construction crew at Wendstedt had a new section to install adjacent to the existing line-up of equipment. Maeve, the construction supervisor, knew of the problem with the extruder. She knew that the extruder motor was being supplied by the alternate source. She also knew that the end plate on the equipment line-up had to be removed for the new section to be installed. Maeve asked Jack if she could install the new section of equipment while the extruder motor was on alternate supply. Jack indicated that would be fine and that it was probably a good time for that work Jack's electricians had the circuit breaker out of its compartment for repair.

On Thursday morning, Maeve assigned Colin and Art the job of adding the new section of equipment. Colin and Art installed locks and tags that prohibited the reinsertion of the primary circuit breaker into its compartment.

Working together, they were about to position the new section adjacent to the existing equipment. It was then that Art noted that the end plate had to be removed from the existing equipment in order to bolt the new section into place. Colin proceeded to remove the bolts and the end plate. Together with Art, he pushed the new equipment into place. The bolts that were supplied with the new equipment were too small for the captive nuts that were in place. After retrieving new captive nuts, the two electricians proceeded to replace the larger captive hardware. Colin was in the back of the cabinet, and Art was in the front.

Colin had installed two of the new nuts, but the third was being difficult. He reached for his spanner to help with leverage. His first action with the spanner was to reach inside the equipment with the tool. Nothing was close except the grounding conductor. When the spanner touched the grounding conductor accidentally, Colin was electrocuted almost instantly.

The grounding conductor was energized, because the motor was running from an alternate source. The "standard" earthing conductor had become the fatal component.

It is absolutely critical that "standard" features be critically evaluated and not simply accepted on blind trust. A standard safety feature can sometimes become a hazard. Understanding requirements in a standard is necessary to continue to avoid injury.

One final thought: Jack had forgotten to tell Maeve that the extruder would normally be down for annual maintenance the very next week. Then it would have been safe to install the new equipment.

Test Your Thinking

1. The essential focus of the *National Electrical Code* is
   a. To define how equipment should be installed
   b. To define how a facility should be designed
   c. To define work practice requirements for electricians
2. The basic physics of how people are injured varies
   a. From one industry to another
   b. From one discipline to another
   c. If a motor vehicle is involved
   d. The physics of how people are injured does not change.
3. International standards attempt to reach consensus among people who represent
   a. European interests
   b. Asian interests
   c. North American interests
   d. Interests across governmental boundaries
4. The system of consensus standards in the United States is
   a. Mandated by state governments
   b. Mandated by the U.S. federal government
   c. Defined by the American National Standards Institute
   d. Voluntary

Answers: 1. (a), 2. (d), 3. (d), 4. (d)

Notes

1. The OSH Act: Public Law 91-596, 1970.
2. This account is based on an actual incident. The names, including the name of the facility, have all been changed to protect those involved. Any similarity to actual names or facilities is strictly coincidental.

■ Chapter 8

Impacting Work Practices

Determining the Work Environment

The work environment and work conditions are almost always significantly different among sites and employers. This difference exists not only among employers but also among sites, product lines, and disciplines of the same employer.

In most cases, electrical energy provides some of the energy necessary to run motors, heat fluids, and provide light. In manufacturing processes, such as an electrochemical process, the electrical energy may be considered to be one element of the process in much the same way that steam may be an integral part of another process. In the metals industry, electrical energy may be the only process "fluid" used. In still other instances, such as the utility industry, electrical energy is the product. Because electrical energy is used in many different ways, it follows that people are exposed to electrical hazards in different ways.

An employee whose assignment is to operate mechanical equipment to extrude a material onto a moving belt may be exposed only to a touch-potential hazard. An employee at the same location who is assigned to troubleshoot the extruder when it fails to run continuously may be exposed to shock, arc flash, and blast while trying to locate the problem.

In an electrochemical process, although the electrical hazards are the same as in the above examples, the degrees of the hazards differ substantially. The type of potential hazard exposure is also substantially different. In an electrochemical process, the process fluid is at an elevated potential. Sometimes the energy is *supplied to* the process, sometimes energy is *supplied from* the process, and sometimes the energy *is* the process.

An employee operating a "cell" producing chlorine might be exposed to potential from a few volts dc to several hundred volts dc. This employee is potentially exposed to a blast hazard as well as a chemical hazard. The degree of potential exposure depends very heavily on the employee selecting the appropriate work practice.

A utility industry employee assigned to record meter readings within a substation fence might be exposed to arc blast, with little or no exposure to a shock hazard. Still another utility industry employee within the operating plant might be exposed to the same type and degree of hazards as chemical industry employees.

Understanding the Work Environment

Effective safe work practices are based on the *type* of hazard and the *manner of exposure,* as well as the *degree of exposure.* In the above examples, an effective safe work practice might require ungrounded tools. In another example, an ungrounded tool might be unsafe. A utility industry safe work practice might be very unsafe if implemented on a metals industry site.

The "process" is an important element of the work environment and must be considered.

When equipment is initially installed and production started, the equipment begins to deteriorate and degrade. All equipment has a life expectancy. As equipment design changes and protective characteristics improve, the method and degree of exposure to an electrical hazard also change. With new equipment, the degree of exposure might be greater than with older equipment, but in all likelihood, the degree of exposure is less.

Necessary troubleshooting is considered a normal condition. Circuit breakers trip; fuses blow; components fail; filters must be changed; overload devices need to be reset. The maintenance program should account for normal actions. Experience has shown that the following situations are possible and even probable:

- Covers and guards might be misplaced and not reinstalled when a worker is trying to get a production line back into service.
- An overload element might be oversized in order to avoid "nuisance" trips.
- Some fasteners might not be reinstalled or latched on doors or covers.

- A "temporary jumper" might be installed.
- A mechanical component might be "blocked" temporarily.
- Corrosion may be significant and cause internal components or enclosures to abridge the integrity of the initial equipment.

People expect the integrity of electrical equipment to be high. Missing or unlatched fasteners are effectively invisible. People expect installed circuit breakers or fuses to be correctly sized and installed. They expect electrical faults to remain contained within the electrical enclosure. People expect grounding conductors and electrodes to remain in place and remain functional. The purpose of the maintenance program should be to make these expectations real: Equipment is closed, overcurrent protection is installed as designed, enclosures are not corroded, and grounding conductors are complete.

The state of equipment maintenance is one important element of the work environment.

The quantity and quality of maintenance tends either to extend or shorten the life of electrical equipment. Executing maintenance tasks generally results in increased exposure to electrical hazards. It is during these activities that enclosures might be abridged while the equipment is energized. Experience has shown that maintenance technicians normally believe in their capability to execute certain maintenance tasks. Checking torque and vacuuming dust from a compartment while the line side of the circuit breaker or switch remains energized are relatively common practices.

Organizational priorities and principles tend to affect such practices. The existence and condition of the employer or site's general safety program has a bearing on the condition of the electrical safety program. In a capitalistic economy, most products are intended to make money for the investors. Therefore, it is normal and necessary for supervisors and managers to be interested in and demand productivity. However, it is important that the supervisory zeal to keep products flowing does not overshadow emphasis on safe work practices. Employees tend to visualize corporate priorities based on how supervisors and managers act instead of what is said or written.

Experience has shown that an organized work site results in fewer incidents and improved productivity. Clutter not only causes distractions, it may cause important information to get lost. Supervisors and workers who are tolerant of cluttered work areas also tend to be tolerant of "cluttered" work practices, resulting in increased exposure to hazards. As exposure increases, so does the number of incidents and injuries.

Personal policies and priorities of people within the organization are an important element of the work environment.

Some personal employee characteristics have a significant bearing on the work environment. Among those characteristics are state of training, age, and experience. These traits generally help define who a person is. Some other traits might change from one day to another. An argument with a spouse or close friend tends to consume a person's thoughts. As the person performs assigned tasks during the day, he or she is less likely to recognize an electrical hazard. Supervisors should select only mentally alert people for tasks where exposure to electrical hazards might exist.

Personal characteristics and priorities of employees are important elements of the work environment.

Weather and climatic conditions tend to affect how a person acts. If a person is uncomfortable due to weather conditions or any other reason, that feeling is likely to influence his or her ability to think clearly. Bulky or uncomfortable clothing is likely to affect the capability of a person to move with confidence. Increased bulk might initiate an incident. Rain, ice, mud, and snow are important factors for a person to consider when selecting work practices.

Physical environmental conditions are an important element of the work environment.

Modifying the Work Environment

Supervisors and managers may affect many important elements that define the work environment. With the exception of weather conditions, members of the line organization can influence each element. Stress is frequently the result of collegial pressure from coworkers, either real or imagined. A person might accept some hazard exposure to avoid admitting being concerned about the exposure.

Communications systems that can impact this human reluctance to admit fear or concern should be put in place.

The most effective communication with employees is nonverbal. In many situations, neither the sender nor the receiver normally recognizes that a message was sent or received. Just as a child observes and emulates a parent or other trusted adult, an employee emulates a leader. Sometimes the leader is a supervisor or manager; at other times, the leader is an older or "wiser" coworker.

Open, effective, and continuing communication with and among employees and members of the management structure significantly influences the work environment.

Training programs generally include accurate and effective information about the technical aspects of electrical equipment and conductors. Some training programs offer detailed information about rules and regulations. However, most training efforts do not cover the process of communicating.

The content of training programs has a bearing on the work environment.

The specific tasks required for effective maintenance vary by equipment type and manufacturer. Consensus standards written for application within specific industries are effective. Maintenance recommendations written by manufacturers covering their particular equipment and national consensus standards written to cover specific types of equipment also are effective. For example, NFPA 70B,[1] *Recommended Practice for Electrical Equipment Maintenance,* 1998 edition, covers maintenance of electrical equipment. It is comprehensive in nature and written by the NFPA committee process. As such, although the document is a recommended practice, it can be viewed in the same vein as a national standard. NFPA 70B is intended to identify practices that provide for continuing service by the equipment. Equipment maintained in accordance with NFPA 70B should be viewed as safe when operating normally, just as a facility installed in accordance with NFPA 70 [*National Electrical Code*® (*NEC*)] should be viewed as safe when operating normally. However, it is noted that safe work practices are not the primary objective of either of these documents.

An employer has a responsibility, which is frequently unrecognized, to embrace standards to be applied at each site. The employer then has a continuing responsibility to ensure that selected standards are appropriately applied. Audits and inspections are effective processes for this purpose. Employees then have a realistic chance to develop an expectation related to equipment and circuit condition when selecting work practices.

Accepting and effectively applying this responsibility impacts the work environment in a positive way.

Comparing 70E with OSHA Requirements: An Analysis of OSHA Standards

Objective, Scope, and Charter

The objective of both NFPA 70E, *Standard for Electrical Safety Requirements for Employee Workplaces,* 2000 edition, and OSHA electrical standards is essentially the same. However, the two organizations are chartered in vastly different ways. The difference in their charter results in fundamental boundary differences. For example, because OSHA exists as a result of U.S. Congressional action, OSHA standards have limitations assigned through the federal statute. However, because the U.S. standards system is voluntary, NFPA 70E is published as a voluntary standard. (Although voluntary, there are legal implications—see Chapter 13, "Safe Work Practices.") The scope of NFPA 70E aligns with the scope of the *NEC*. The OSH Act provides authority for the Secretary of Labor to write standards applicable to employers. OSHA rules apply to employers. NFPA has no such limitation. NFPA 70E, then, can include directions for employees as well. OSHA can only suggest that employers and employees work together; NFPA can define such a requirement.

Four OSHA standards cover electrical safety-related work practices, in addition to lockout/tagout. NPFA 70E addresses all electrical safety-related work practices in one document. Although NFPA 70E is relatively complete, some exceptions remain. The scope of the document is the same as the *NEC;* therefore, the utility industry is exempted. All other industries are intended to be included within the scope.

Construction Industry

OSHA is subject to political pressures. These pressures tend to drive OSHA standards to be specific for a single discipline. As a result of these pressures, hazards associated with electrical energy are covered in many different places in the OSHA rules. The rules in Title 29, Part

1926, all apply to the construction industry. OSHA makes an attempt to address construction electrical hazards in Subpart K—"Electrical," and in Subpart V—"Power Transmission and Distribution." Subpart K seems to be based on an understanding that employees engaged in "construction" are not really exposed to electrical energy hazards. However, electrical hazards exist in every instance where electricity is used, even in construction tasks.

Federal statutes are complex. Several federal laws existed prior to the OSH Act (e.g., the Davis-Bacon Act and the Contract Work Hours and Safety Standards Act). The term *construction* is defined as "work for construction, alteration, and/or repair, including painting and decorating."[1] In Part 1926, OSHA references this definition, essentially embracing the definition for the construction standards. Although this definition leaves much undefined, it does offer an insight into the intent of the term *construction*. From the OSHA standards point of view, construction should be considered as tasks instead of occupation.

One viable way to gain understanding of OSHA terms is through litigation. Courts do not always agree. However, with sufficient judgments, it is possible to begin to understand how the legal community might define certain terms. These judgments tend to define construction as installing something new that did not exist before the activity. Any activity that replaces a component with a like component is maintenance. If the component did not exist or if the new component is of a different fabrication, then the activity is probably construction. The content of Subpart K covers construction as opposed to either contractors or construction employers.

In the purest sense, the amount of electrical energy supplied for construction work is small. Electricity operates only lighting and hand tools. Normally, electrical energy is contained within enclosures just as it is in a manufacturing plant. The construction work involves selecting and installing concrete, steel, conduit, and wiring, along with all the other components required to complete the facility. Exposure to electrical hazards is, therefore, also limited. Exposure to electrical shock is the prime hazard. However, the degree of exposure is elevated because the temporary wiring is likely to be a class less than required for the completed facility. (See Article 305 of the *NEC.*)

Maintenance of electrical equipment installed to provide temporary electrical energy for construction work, such as a power distribution panel, does not fit the definition of construction described above. Maintenance of temporary electrical facilities, then, must be covered by another section of the OSHA standards.

OSHA standards have several base references. As indicated earlier, the construction standards found in 29 *CFR* 1926 incorporate preexisting U.S. regulations. One of these regulations is the Construction Safety Act, a federal regulation that predates OSHA. Much of the current construction standards came from this act.

Construction equipment that requires motion or movement generally receives energy from a gasoline- or diesel-powered engine. The energy conversion for heavy equipment is from mechanical energy to mechanical energy instead of electrical energy to mechanical energy. The construction standards, then, are aimed primarily at hazards usually associated with construction work, such as falls and falling objects, earth cave-ins, and fires.

Therefore, requirements for construction standards are vastly different from those of general industry. For example, in 29 *CFR* 1926.417, OSHA covers lockout/tagout for construction. In this section, OSHA permits control devices to be tagged. There are no requirements for locks or lockout devices in 29 *CFR* 1926.417, an apparent contradiction when compared with the requirements in 29 *CFR* 1910.147.

Section 1926.416 reads: "When fuses are installed or removed with one or both terminals energized, special tools insulated for the voltage shall be used." This text seems to suggest that it is acceptable to exchange fuses with the terminals energized. Fuses should never be exchanged when the terminals (either line or load) are energized when it can be avoided. Exchanging fuses with a "hot stick" is a normal practice, but, even in this circumstance, the fuse should not be energized until it is again "closed into" the circuit.

29 *CFR* 1926 (Subpart K)—1926.400–449—"Electrical"

Part 1926, Subpart K, is inadequate. This standard offers little, if any, protection from hazards associated with electrical energy. The only protective value of Subpart K—found in Part 1—is a requirement to provide ground-fault circuit interrupters (GFCI). Of course, even the GFCI requirement is included in later editions of the *NEC*. Regardless of its basis, Subpart K should be eliminated and replaced with a standard, such as NFPA 70E, that adequately addresses electrical hazards. Subpart K does not help prevent injury where the *NEC* is adequately applied.

29 *CFR* 1926 (Subpart V)—1926.950–960—"Power Transmission and Distribution"

On the other hand, Part 1926, Subpart V, is very closely aligned with requirements for "general industry" in 29 *CFR* 1910.269 (Subpart R). This section of the "construction" standards contains the same acceptable and unacceptable practices. Unlike Subpart K, Subpart V does support the fact that employees are exposed to electrical hazards while executing "construction." This section relies on a few work practices to provide protection for employees, including using insulated gloves, tagging, and grounding. Subpart V is effective; however, the attempt to compartmentalize the OSHA rules for an industry detracts from the objective to prevent injury.

General Industry

29 *CFR* 1910.269 in Subpart R—"Electric Power Generation, Transmission, and Distribution"

[29 *CFR* 1910.261–275 is Subpart R. Section 1910.269 is "Electric power generation, transmission, and distribution."] Even in the general industry section of the OSHA rules, considerable pressure exists for OSHA to write standards for an industry or industry segment. Of course, the intent of the industry association lobbying for an industry standard is legitimate, at least as viewed by the association. The standard covering the utility industry, 29 *CFR* 1910.269, is an example, although little lobbying was necessary.

The laws of physics governing electrical energy do not change from one industry to another. As a result, hazards associated with electrical energy remain the same from one industry to another. However, the degree of exposure to electrical hazards may change among industries and industrial segments. Again, the utility industry is a good example. The generally higher voltages present in the utility industry result in higher current flow in the event of an accidental contact. Higher fault-current capacity results in more significant blast pressures in the event of an accidental release. Generally, the need to provide continuous service results in longer arcs in the event of an accidental energy release.

Again, however, Section 1910.269 of Subpart R considers generation, transmission, and distribution each as an "activity" instead of an industry segment. Because the utility industry was heavily involved in writing the content of this standard, it is based on utility industry standards and practices. In the same way that the *NEC* is *the* standard for residential, commercial, and general industry installations, the *National Electrical Safety Code (NESC)* governs utility installations. The *NESC* contains some work practices; therefore, the *NESC* serves as one base reference standard for 1910.269.

The title of 29 *CFR* 1910 Subpart R is "Special Industries." The following sections are covered in this subpart:

1910.261   Pulp, paper, and paperboard mills
1910.262   Textiles
1910.263   Bakery equipment
1910.264   Laundry machinery and operations
1910.265   Sawmills
1910.266   Pulpwood logging
1910.267   Agricultural operations
1910.268   Telecommunications
1910.269   Electric power generation, transmission, and distribution
1910.272   Grain handling facilities

Understanding how to apply the standards this subpart comprises is difficult because the subpart title suggests that the content is intended to cover industrial segments. The title of each section also suggests that the intent is to apply the content to the sawmill industry. This same logic would suggest that 1910.269 applies to the electric utility industry.

However, to correctly apply Subpart R, one must recognize that in its legislative charter, OSHA is given the authority to regulate employers, not industries. Instead of industries, then, each section of Subpart R applies to special activities or tasks. In other words, the term *generation, transmission, and distribution* really means facilities, installations, or services that fulfill the activity or task of generating, transmitting, or distributing electrical energy. This point is key because it effectively means that almost every industrial facility has at least one installation to which this standard applies. Most industrial facilities have installations that are intended to be covered by this standard.

29 *CFR* 1910 Subpart S—1910.301–399—"Electrical"

When the current safety-related work practices standard was promulgated in the *Federal Register* on August 6, 1990, the preamble indicated that NFPA 70E content served as the base standard on which Subpart S was predicated. The section was organized (and remains) in the same form as the reference standard. The subpart was divided into the following four parts:

- Design safety standards for electrical systems. These regulations are found in 1910.302 through 1910.330.
- Safety-related work practices. These regulations are found in 1910.331 through 1910.360.
- Safety-related maintenance requirements. These regulations are found in 1910.361 through 1910.380.
- Safety requirements for special equipment. These regulations are found in 1910.381 through 1910.398.

These subdivisions parallel the sections of NFPA 70E. Part I of the 1985 edition of NFPA 70E was adopted verbatim and placed in the OSHA standard as 1910.302 through 1910.308. Prior to this time, OSHA had relied on the 1972 edition of the *NEC* to define the requirements for general industry. OSHA determined that Part II of the 1985 edition of NFPA 70E contained language that was not enforceable and, therefore, could not be adopted verbatim. Using NFPA 70E as a guide, OSHA authored and promulgated the text currently contained in 1910.331 through 1910.335.

Although the initial committee completed Part III of the 1985 edition of NFPA 70E, this section of the document contained no mandatory requirements. Consequently, Subpart S contains no safety-related maintenance requirements.

The safe work practice sections (1910.331–335) of Subpart S address all issues mentioned in the 1985 edition of NFPA 70E. The OSHA standard does a good job of defining requirements to avoid some exposure to shock. In 1910.333, the following statement seems to suggest that other hazards might also exist.

> (a) General. Safety-related work practices shall be employed to prevent electric shock or other injuries resulting from either direct or indirect electrical contacts, when work is performed near or on equipment or circuits which are or may be energized. The specific safety-related work practices shall be consistent with the nature and extent of the associated electrical hazards."[2]

This statement seems to suggest that shock may not be the only electrical hazard. However, Subpart S contains no requirements related to any other hazard. It is certain that hazards discussed in Chapter 2 of this book exist and result in injuries to people. This means the employees in an organization complying with all Subpart S requirements may still be exposed to arc-flash and arc-blast hazards. The 1985 edition of NFPA 70E also only addressed shock as a hazard. Industrial experience has shown that shock injuries comprise about 50 percent of all injuries related to electrical hazards.[3] Assuming that full compliance with Subpart S requirements *and* assuming that Subpart S requirements would prevent all shock injuries, electrical injuries would be reduced only by one-half.

When 29 *CFR* 1910.269 was promulgated, OSHA established a minimum clothing requirement for people exposed to arc flash. However, the requirement is found only in the safe work practices in the utility industry. Of course, arc flash is a significant hazard in the utility industry and in general industry as well. Newspapers frequently report that an electrical explosion has occurred, with resulting injuries. Although Subpart S does not ignore exposure to an arc-flash hazard in general industry, this type of injury continues to occur.

Section 1910.147 in Subpart J—"The Control of Hazardous Energy (Lockout/Tagout)"

[29 *CFR* 1910.141–150 is Subpart J. Section 1910.147 is "The control of hazardous energy (lockout/tagout)."] Section 1910.147 is the prime section covering lockout and tagout. However, lockout and/or tagout are discussed in other places in OSHA standards. For example, the electrical safety-related work practices for both power generation, transmission, and distribution [1910.269(d)] and general industry [1910.333(b)(2)] include requirements for lockout or tagout. (Chapter 9 of this book is devoted entirely to lockout/tagout.)

 ## A Closer Look

Lois and Dick were contractor employees, assigned to a textile manufacturing plant.[4] They were part of a crew of contractor employees assigned to install a new motor control center (MCC) during a plant turnaround. At the plant, the turnaround was the period of time in which annual maintenance was performed. The new motor control center was not required for the plant to restart, but terminations were necessary while the unit substation could be shut down. With this plan, the textile plant could avoid working on energized bus in the future. This was a good idea that would avoid exposure to shock in the future.

Lois and Dick had the new MCC set into place prior to the shutdown, with cable tray and conduit already installed. Load cables and utilization equipment would be added in the future, as new equipment was needed for the new product.

Statistically, the textile plant had a good safety record. Statistics suggested that personal safety was highly valued by the organization. The contract between the textile plant and the contractor was standard. The contractor was responsible for testing, as necessary, to ensure the integrity of the installation. The contractor was to follow all legal requirements, including OSHA requirements. However, the contract failed to indicate which portion of the OSHA standards contract employees were expected to follow.

Lois and Dick were familiar with all requirements of the construction standard in 29 *CFR* 1926. In fact, their last training had relied on the pamphlet obtained from OSHA that included all construction standards. They knew when safety harnesses were needed. They were familiar with shoring requirements. They knew about GFCIs and used one on all electrical hand tools. Their employer was a licensed construction contractor. They were proud of their abilities and careers.

On Monday of the shutdown week, Lois and Dick pushed the feed cable into the main feeder section of the MCC, terminated the conductors, and marked each conductor with red, blue, and yellow identification tape, according to the plant standard marking system. Next, they proceeded to the secondary switchgear and pushed the conductors into the correct compartment. Robert, the textile plant electrician, told Lois and Dick that the unit substation was deenergized. He showed them the unit substation main switch so they could hang a tag, if needed. Until now, no tags had been necessary.

Of course, 29 *CFR* 1910, Subpart S, contained the OSHA rules that guided Robert as he executed his maintenance and servicing work. Like Lois and Dick, Robert was highly respected for his knowledge and ability. He was well schooled in the electrical safety-related work practices defined in Subpart S.

After hanging a tag on the main switch, Lois and Dick proceeded to train the new conductors into position and terminate them on the load side of the spare breaker. They considered their work complete and told Robert they would see him later.

The construction supervisor asked Lois if the new cable and MCC had been tested. Lois indicated that the cable had been tested immediately after it was pulled, and she gave the supervisor the record of that test. The supervisor assigned the two electricians to other work for the rest of the afternoon and delivered the cable test record to Robert.

The next morning, the contractor supervisor asked specifically if the MCC had been tested. Lois indicated that the MCC had been tested when it was installed, but no record was kept. Noting that a significant amount of time had passed since the MCC was set into place, the supervisor asked Lois and Dick to test the MCC structure before they released the new unit to the textile plant for use.

Early Thursday morning, Lois and Dick checked out the high-potential (hi-pot) equipment from the shop and went to test the MCC structure. In order to isolate the MCC for the test, Dick removed the conductors from the main feeder section. Even though the conductor ends were touching, there was no need to tape the conductor ends. After all, they would be reterminated as soon as the test was completed. The door of the main compartment was left open so they could connect the hi-pot leads. Neither Lois nor Dick recognized that their work was no longer "construction work." This last task should have been executed as servicing and maintenance work, because the MCC had been connected to a source of energy.

In the textile plant electrical shop, Robert had completed all his maintenance jobs and was preparing for the plant to start production on Friday morning, as planned. He elected to go to the same substation and heat it up.

The secondary switchgear was now energized. Next, Robert proceeded to energize MCC units and other equipment supplied by the substation.

Lois had connected the hi-pot leads and was kneeling next to Dick to operate the machine when Robert energized the new feeder. A violent explosion occurred. Most of the energy went over the heads of both Lois and Dick. However, both were sprayed with droplets of molten copper. Neither Lois nor Dick was physically injured to any significant degree, but both strained several muscles as they attempted to scramble out of the way. The door of the compartment scraped and cut Lois's back as she tried to escape.

The circuit breaker in the secondary switchgear did its job and rapidly opened the circuit. Its proper operation prevented a more significant fireball.

Robert believed that the work was complete. Lois and Dick thought the startup was scheduled for Friday morning, and they did not expect the unit to be energized.

The textile plant electrician and the construction electricians were following different rules. Each facility should have had a set of rules that applied to everyone on site. Following different standards almost always sets a trap that likely will spring at an unexpected time.

 ## Test Your Thinking

1. Effective safe work practices are based on which of the following:
   a. Type of hazard
   b. Manner of exposure
   c. Degree of exposure
   d. All of the above
2. Which of the following standards covers maintenance of electrical equipment?
   a. NFPA 70E
   b. NFPA 70B
   c. NFPA 70
3. Which of the following industries is not covered in 29 *CFR* 1910, Subpart R?
   a. Electric power generation, transmission, and distribution
   b. Telecommunications
   c. Construction
   d. Textiles

4. Which of the following correctly cites an OSHA standard?
   a. 29 *CFR* 1910.147
   b. 1910 *CFR*, Section 147
   c. OSHA 1910, Part 147
   d. Subpart 147, Section 1910 from OSHA

Notes

1. OSHA regulations, 29 *CFR* 1910.12.
2. OSHA Regulations Section 1910.333(a) [Federal Register/ Vol. 55, No 151, Monday, August 6, 1990].
3. R. L. Doughty, R. A. Epperly, and R. A. Jones, *Maintaining Safe Electrical Work Practices in a Competitive Environment.*
4. This account is based on an actual incident. The names, including the name of the facility, have all been changed to protect those involved. Any similarity to actual names or facilities is strictly coincidental.

Answers: 1. (d), 2. (b), 3. (c), 4. (a)

■ Chapter 9

Lockout/Tagout

The only real way to avoid incidents and injuries related to energy is to avoid all exposure to sources of energy. Energy is required for equipment to operate and produce a product. Therefore, preventing incidents and injuries boils down to avoiding exposure to hazards associated with the energy. To avoid electrical injury, a person must avoid exposure to shock, arc flash, and blast (see Chapter 2, "Electrical Hazards"). Electrical equipment and services generally are free from electrical hazard while they are operating normally. The trick, then, is to avoid exposure to these hazards when conditions are *not normal.*

Many consensus standards include a requirement for lockout/ tagout. However, most consensus requirements depend heavily on a series of methodical actions that are learned in training. Methodical thinking, as usually involved in a procedure, and the associated training are very important. In reality, however, avoiding injury requires more than simply following a procedure or rule. Lockout/tagout rules and procedures may be the most misused, misapplied, and least understood of all regulations.

Establishing an Electrically Safe Work Condition

In the case of electrical energy, avoiding exposure is more complex than suggested by lockout/tagout. (This is really true for other energy sources as well.) To avoid exposure to electrical hazards, an electrically safe work condition must be established. Achieving an electrically safe work condition involves application of a six-step process:

1. Determine all sources of energy by reviewing up-to-date drawings.
2. Disconnect all sources of energy by operating adequately rated disconnecting means.

3. Inspect, wherever possible, energy-isolating devices for visible breaks in power conductors.
4. Perform a voltage test to determine the absence of voltage.
5. Install grounding devices, if determined to be necessary.
6. Install locks and tags.

(For a procedure or practice example on creating an electrically safe work condition, see Appendix B.)

When an electrically safe work condition exists, no danger of release of electrical energy exists. All hazards from electrical energy have been removed. Lockout/tagout, as required by OSHA and other consensus standards, however, leaves a significant potential for electrocution or other release of electrical energy.

Determine All Sources of Energy by Reviewing Up-to-Date Drawings

In most cases, this need cannot be met because the drawings are not up-to-date. Changes are frequent in most industries. Commercial facilities generally do not change as drastically, but change in electrical services is still an issue. Frequent changes, coupled with tight money, results in drawings being out-of-date when needed. Single-line diagrams are critical when determining which disconnect to lock out. Other drawings may be substituted for single-line diagrams, provided they clearly illustrate *all* sources of electrical energy.

When checking drawings to locate all sources of energy, analyzing the work on the basis of its electrical location in the circuit is very important. The analysis must also consider any temporary sources of energy or temporary cross connects to other electrical equipment. Checking for temporary conductors is still another aspect of up-to-date drawing information. The technical folks usually are the ones who fail to mark temporary conductors on drawings.

Each installation should adopt a standard identification system. All switches and circuit breakers should be marked to show where the energy comes from and where it goes. Most consensus standards, including the *National Electrical Code®* (*NEC*) and OSHA, require this information.

Both drawings and labels are necessary. Both should be checked, and any discrepancy should be explained satisfactorily. The person-in-charge and all employees should understand the results of checking the drawings and labels.

Disconnect All Sources of Energy by Operating Adequately Rated Disconnecting Means

In some cases, disconnect switches are installed to provide a point for lockout. It is relatively common for a design to install a lockout switch near motors or equipment that are frequently shut down for mechanical cleaning or adjustment. In such cases, the design intent and the operating procedure are for the motor to be stopped with the push button or control device prior to opening the disconnect switch. The disconnect switch is likely inadequate to break the full-load current. Disconnecting devices should be adequately rated to interrupt the full-load current. Switches that are load rated are so labeled by the manufacturer. It is recommended that an additional tag—"For Lockout Only" (in large letters)—be applied on each lockout-only disconnect switch. (See Figure 9–1.)

Inspect, Wherever Possible, Energy-Isolating Devices for Visible Breaks in Power Conductors

Manufactured components begin to deteriorate when they are installed. Devices that are basically mechanical, such as disconnect switches and circuit breakers, eventually will fail. In many cases, the failure mode is that the "blades" are dislodged from the operating handle. The failure can involve one or more of the contacts. The only real way to know if this failure has occurred is to open the door and visually inspect the blades for a physical opening.

In some instances, it is difficult or even impossible to see the physical break. *Purchasers of electrical equipment should insist on having the capability to inspect for a physical break.* In some equipment, the physical break is covered by another component. In some instances, the manufacturer paints or otherwise marks the blades for inspection. Although marking the blades is a positive step, it *does not* decrease the need to view the opening in the power conductors.

Until a safe work condition exists, all conductors must be considered to be energized, regardless of any other circumstance. Arc-flash

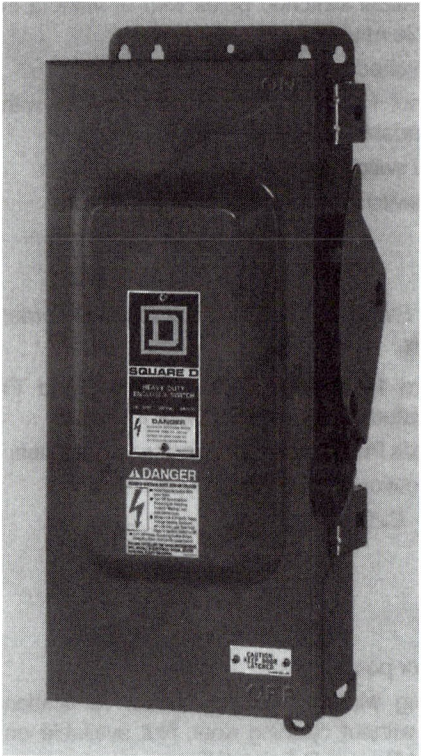

Figure 9–1. A Disconnect Switch.

protection is necessary to perform this inspection, as doors must be opened.

Perform a Voltage Test to Determine the Absence of Voltage

Electrical technicians and engineers usually are very interested in voltage. In most cases, their interest is in how much voltage is present compared with how much voltage is supposed to be there. For lockout purposes, however, the idea is to *verify the absence of any voltage.* In fact, it is absolutely critical that the voltage measurement be complete.

One common problem when making a voltage measurement is to have the meter set to the wrong scale. Another is to have the meter leads inserted into the wrong plugs on the meter case. Using a single-function voltmeter avoids these problems. As a matter of fact, a single-function voltmeter where the leads are not readily changeable avoids both problems.

Like all electrical equipment, voltmeters fail. To ensure that the device is operable, it is wise to verify that the meter is functional on another energized circuit both before and after its use. Because any voltage higher than 49 volts is hazardous, this verification is just as critical for control voltages as with medium-voltage conductors.

Solenoid-type voltmeters generally have a duty-cycle rating. Many people like the feel and sound of a solenoid-type voltmeter. Frequently, these same people are not aware of the implications of the duty-cycle rating. In some cases, these devices have literally exploded in a person's hand by maintaining contact with an energized conductor for too long, therefore exceeding the rated duty cycle.

Personal protective equipment determined when the hazard analysis was performed must be worn until a safe work condition exists.

Install Grounding Devices, If Determined to Be Necessary

Historically, grounding circuit conductors has been a highly valued protective scheme. It remains no less valuable today. Consensus standards and training programs place significant emphasis on temporarily grounding a conductor or set of conductors. To avoid the possibility of induced voltage or the potential for stored energy, temporary grounds are recommended.

Where there is a remote possibility that the conductor could become accidentally reenergized, it is important to establish a zone of equipotential where work will be occurring. Equipotential simply means that safety grounds are installed in such a way (perhaps on both sides of the work activity) to preclude the possibility of a step- or touch-potential hazard. (See Figures 2–2 and 2–3 in Chapter 2, "Electrical Hazards.")

Some potential hazards are associated with installing safety grounds. These hazards are associated with the integrity of both the ground set and the method of installing the set. In many cases, an electrician fabricates a set of grounds for temporary use, thinking that there is no voltage and that the ground is "just in case." He or she might not understand what would happen to the ground if voltage should be reapplied.

In other instances, temporary "safety grounds" are accidentally left in place after the work is complete. Occasionally, the circuit is reenergized with the ground set still in place. Maximum stress is applied to the ground set at this point. If the ground set is rated for the service, the overcurrent device will operate. If the ground set is inadequately rated, the fault will rapidly become an arcing fault.

Install Locks and Tags

Locks should be personal locks and tags should be "danger " tags (see Figure 9–2). Neither the locks nor the tags should be used for any other purpose. Tags should identify who, when, and why the tag was applied. Keys to the lock should be in the possession of the person who installed the lock. The installation of locks and tags should be guided by standardized practice on a site. The person-in-charge should be aware of all locks installed, as well as when the locks are removed. As the number of lockout points increases, the responsibility of the person-in-charge increases.

Only the person who installed the lock should remove it. However, the site procedure must define the process to be followed in the rare case when a person who installed a lock cannot be located to remove it. It must *not* be acceptable to remove another person's lock. Punishment for doing so must be defined, published, understood, and rigidly enforced.

Lockout locks should never be installed without a danger tag. Only in rare circumstances should a danger tag be installed without a lock. Although permitted by most consensus and OSHA standards, tagout (no lock) can never be as secure as lockout (with a lock).

Lockout/Tagout in OSHA

Requirements

In OSHA 29 *CFR* 1910.147, requirements are aimed at eliminating exposure to hazards associated with unexpected energy release. All people performing servicing or maintenance on a system or equipment must apply individual locks and tags. They must be "in control" of the energy. Exposure to electrical hazards is exempted from 1910.147. In concept, 29 *CFR* 1910.333(b)(2) or 1910.269(d) define lockout/tagout requirements for exposure to electrical hazards. It is important to note that, although exposure to electrical hazards is exempted, electrical energy is not. The basic idea of 1910.147 is to con-

Figure 9–2. Danger Tag.

trol the *energy,* thus eliminating potential exposure. In the electrical standards, the basic idea is to control the *exposure,* while accepting exposure by qualified employees.

Lockout/tagout concepts found in 1910.147 and embraced in the electrical standards include the following:

- Applies to servicing and maintenance activities.
- Specifies one person–one lock.
- Permits tagout.
- States preference for lockout over tagout.
- Emphasizes training.
- Exempts direct exposure to electrical hazards.
- Specifies written program.
- Promotes audits of program effectiveness.

Application to Servicing and Maintenance

A key point is that Section 29 *CFR* 1910.147 applies only to servicing and maintenance. In the OSHA instruction to inspectors, the following advisory makes that point: "The new rule addresses practices and procedures that are necessary to disable machinery or equipment and to prevent the release of potentially hazardous energy while maintenance and servicing activities are being performed."[1]

In 1919.147, the OSHA standard uses the terms *servicing and maintenance* and *unexpected release of energy*. When equipment is functioning normally, of course, locks and/or tags are not an issue. The standard suggests that controlling energy or exposure to energy is only an issue when equipment is either being repaired or maintained. When in normal operation, equipment operators and maintenance people know when and how they are or may be exposed to electrical energy. In fact, exposure to electrical energy in normal conditions should not be possible. Lockout or tagout becomes an issue during "not normal" conditions.

Understanding the meaning of "servicing and maintenance" requires some study of text found in the preamble to 1910.147 and other supporting information, such as the OSHA Inspection Instruction STD 1.73. After studying these information sources, it is clear that the rule is not intended to apply to normal operations. In the OSHA standards, lockout/tagout attempts to offer protection for employees repairing failed equipment, performing normal maintenance, or correcting abnormal equipment operation.

In 1910.147, a second point is the understanding of the term *unexpected release of energy*. Again, from the study of the preamble and the inspection instruction, the application of the term is intended to broaden the meaning of servicing and maintenance. In the production process, the use of robots and conveyors is common. Machines that eject a product to move along a conveying system sometimes become jammed by the random nature of the physical movement. In other instances, products or raw materials are handled in open movement by mechanical arms or compressed air or otherwise physically manipulated. Again, the unwanted randomness of movement might cause the product to become misaligned on the conveyor belt. Correcting such a happening might not be considered "servicing and maintenance"; therefore, lockout may not be required.

One Person–One Lock

The concept of one person–one lock is a good one, especially where a small number of disconnect switches or circuit breakers are involved and where the energy supply originates. However, in many instances electricians are not familiar with the location and identification of the sources of electrical energy. In the case of a plant turnaround, employees frequently are transferred from one plant or area to another to help with the work. Sometimes contract employees are brought in to supplement the numbers of internal employees.

It is unrealistic to expect that each person understand where locks should be installed to avoid exposure to electrical hazards. Equipment operators or normally assigned maintenance employees likely will know where locks and tags are needed. However, experience has shown that, in reality, the only effective method is to identify a "person-in-charge," as defined in NFPA 70E, *Standard for Electrical Safety Requirements for Employee Workplaces,* 2000 edition, Part 2, Chapter 5. The person-in-charge is assigned the responsibility to make certain that all energy control points are opened and effectively controlled by locks. It is equally important that each person who is potentially exposed to electrical energy hazards has some assurance that the energy will not be released while he or she is exposed. Locks and tags are required to be installed and removed by the same employee. Even in the case of multiple crews and employers, the "person-in-charge" idea is more effective when applied in conjunction with the one person–one lock concept.

Use of Tagout

Tagout has been used effectively in the utility industry for many years. Transmission and distribution tagout processes rely on a dispatcher installing tags to protect other employees. The wide spacing between isolating points and isolating methods—many miles in some instances—determines the need for a concept other than one person–one lock.

Under certain conditions, tagout is acceptable by OSHA standards in general industry. During the public comment period for 29 *CFR* 1910.147, some industry groups lobbied strongly to use only tags for energy control. Although tagout was included in the promulgated standard as a viable protection scheme, a condition was included that an employer must prove that tagout is as effective as lockout if tagout is used. The same requirement is included in NFPA 70E. Although intellectually possible, establishing this proof is difficult. In both NFPA 70E and OSHA standards, if tagout is used by an employer, then at least one additional step must be taken before operating the isolating device. Clearly, lockout is preferred to tagout.

Importance of Training

Lockout/tagout training is very important. Training provided to employees should be developed specifically for an employer and for a particular site. The training might be acceptable for more than one site; however, each site should be analyzed as a separate entity. Training must be based on the site or employer lockout/tagout procedure, which in turn, must be based on OSHA and consensus standards.

Direct Exposure to Electrical Hazards

Knowledge, ability, and skill play important roles where exposure to electrical hazards exists. These characteristics are generally accepted within the electrical discipline as primary protection for qualified employees working near an energized conductor. OSHA does permit work directly on an energized point in some circumstances.

Although working directly on an energized conductor is restricted, working *near* an energized conductor is much less restricted. Industrial experience suggests that working *near* an exposed energized conductor is more likely to result in injury than working directly *on* an energized conductor. When a person is working directly on an energized conductor, his or her expectations are more focused. When working near an energized conductor, maintaining a conscious awareness of the electrical energy is much more difficult.

Testing the circuit before contacting it is absolutely necessary to avoid injury. In fact, if the voltage is more than 600, the test instrument operation must be verified immediately after performing the test [required in OSHA Subpart S, 29 *CFR* 1910.333, (b)(2)(iv)(B)]. There is no logical explanation of why it is important to test equipment operation at more than 600 volts and not at voltages less than 600. This requirement almost seems to suggest that if the voltage is less than 600, electrical hazards are not that significant. This is certainly not the case. In fact, most industrial electrocutions occur at voltages less than 600. Verifying that the voltage test is complete and accurate has prevented many electrocutions. The OSHA standard is inadequate on this point. NFPA 70E requires that instruments used to perform a test for absence of voltage be tested for proper operation both before and after use. This procedure provides the best chance of a valid test.

Another point that must be made about voltage testers is that the market is flooded with inadequate test devices. Any person who must test for voltage at any time must ensure that the voltage tester he or she uses is adequate for the task, based on the capacity of the circuit. A voltage rating is established by the manufacturer. However, verification of the rating by an independent laboratory provides the best chance that the rating is adequate.

Written Program/Audits

A procedure cannot be applied effectively unless it is written and available to all employees. The program must be readily available and continuously reviewed for possible improvement. In order to determine possible procedure improvements, regular audits with written records must consider four important elements of the program:

1. Does the procedure cover all instances where potential exposure to hazards associated with energy exists?
2. Do the employees apply the energy control program effectively?
3. Has the equipment or service been changed since the employees were trained?
4. Have all employees received training on the procedure?

(For a sample lockout/tagout procedure, see Appendix B.)

 A Closer Look

On Thursday night, Kevin was working the second shift at Rouse Pharmaceutical,[2] a small prescription drug-manufacturing facility. Few maintenance people were required to keep such a small facility running. In fact, the entire electrical maintenance force consisted of only four people. In order to provide maintenance coverage around the clock when one person was on vacation, extending the shift work to twelve hours provided primary electrical maintenance. Kevin had drawn the second shift this Thursday night.

Kevin had worked at Rouse for 17 years and was considered to be the best electrician on the site. He had never had an injury in his career as a maintenance electrician. Kevin was the person whose opinion was always sought in "sticky" situations. He was enthusiastic and skilled. Like most Rouse employees, Kevin was acutely aware of the need to keep costs to a minimum. Consequently, he routinely looked for productive work instead of waiting for a trouble call.

It was difficult to see in some areas of the plant because of shadows created by process piping that was added and changed through the years. The plant manager had authorized a project for additional lighting, and the work had been started at the beginning of the week. Locations for several light fixtures had been selected; some new lights were already in use.

On this night, all planned maintenance work was completed before the meal break at midnight. As he was eating his food, Kevin considered what he could do to be productive during the early morning hours. He remembered the lighting project and decided that he would hang a couple of fixtures. After completing his meal, Kevin told the shift supervisor of his plan to install the new light fixtures. Knowing of no other productive electrical work, the supervisor agreed with Kevin's idea.

The batch-mixing area was one place where more light was needed. It was about 1:00 A.M., when Kevin found the new fluorescent fixture and started the work. He was working alone, because he was the only electrician in the plant. One of the new fixtures was to be located near a walkway and below new piping. The obstructing piping was adjacent to the platform, preventing ladder access to the ceiling. By standing on the second rung of the handrail, Kevin could reach the location from the platform.

Conduit had been installed from the new fixture to a junction box serving the existing area lighting. Kevin proceeded to hang the new fixture on chain and installed a short piece of flexible conduit to the fixture. After a break at 2:30 A.M., Kevin checked with the shift supervisor to see if anything needed his attention. Nothing did, and Kevin returned to the lighting job.

After gathering the necessary material, Kevin climbed the stairs on the opposite end of the room and made his way back to the new fixture. The platform on which he was working was on the opposite end of the room from the stairs. The lighting panel was located in an adjacent room, near the stairway. The ceiling area in batch mixing was congested. The access walkway had many obstructions that had to be negotiated in order to reach the work location. In fact, the walkway was really a series of connected platforms.

After reaching the location, Kevin proceeded to install the conductors into the new conduit and to connect the wires to the fixture. The area lighting was needed for visibility. Had the lights been turned off, temporary lights would have been necessary for Kevin to see what he was doing. The lights were not needed for plant production, because only one product batch had to be mixed per shift. The day shift would need lights, but Kevin would be finished long before then.

Kevin never really considered deenergizing the circuit. He really did not want to work by temporary or hand lighting; anyway, it would take him another half hour to find the lights, run extension cords, and so on. It would be difficult for him to traverse the walkway without decent lighting. He was confident of his skill and ability. After all, the circuit was only 110 volts, and he had worked that voltage "hot" for his entire career.

The plant had no electrical safety program. The employees had no policies and no procedures. The electrical crew could readily determine what needed to be done and how to do it.

It must have been about 3:15 A.M. when Kevin attempted to connect the new light fixture in the junction box. He must have been standing on the middle rail of the walkway to reach the junction box. However, Kevin never completed the work.

About 5:00 A.M., when the shift was almost over, the shift supervisor came into the batch-mixing area to check vessel levels. When he entered the room, he had a feeling that something was wrong. He saw a crimping tool on the floor in front of the door he had entered. When the shift supervisor looked up, he saw Kevin lying suspended on the maze of pipes above. The supervisor called to Kevin, but there was no response. The supervisor used the nearby phone and called for emergency help. The emergency crew responded and removed Kevin's lifeless body.

An autopsy determined that Kevin had been electrocuted. A small burn was found on the forefinger of one hand. No other burns were apparent.

In the junction box, splices in the grounding and neutral conductors were complete. These were the first two of the three connections that Kevin had to make. The wire nut on the "hot" conductor had an indentation that penetrated the insulating material. The crimping tool found on the floor below had a small burn that matched the burn on Kevin's forefinger.

It appeared that Kevin had trouble gripping the wire nut tightly enough to remove it from the existing connection. In order to gain a better grip, he apparently had gripped with too much force and penetrated the insulating layer. His hand was near the fulcrum point of the crimping tool, with his forefinger in contact with the tool. Kevin's legs were in contact with the handrail at the same time his hand was in contact with the tool, and the tool was in contact with the energized 110-volt conductor.

Kevin had made a fatal mistake by working on a "live" circuit. His experience told him that working with live 110-volt conductors was not really a problem. There was no real reason to lock out the circuit. He tried to save the company a few bucks and himself a few steps.

It would have been simple for Kevin to establish an electrically safe work condition, but he did not. There was valid reason to deenergize the circuit and lock it out, but he did not. Kevin should not have used his crimping tool as an extractor, but he did. The plant could have had a lockout procedure, but it didn't. Rouse Pharmaceutical could have had a policy against working on "live" conductors, but it didn't.

Kevin wouldn't attend his son's baseball game on Saturday—neither this Saturday nor any other.

Test Your Thinking

True　　False

☐　　☐　　1. Establishing an electrically safe work condition is a six-step process.

☐　　☐　　2. If the person who installed a lock on a piece of equipment cannot be found, it is acceptable for his or her supervisor to remove the lock only if defined in the site procedure.

☐　　☐　　3. Both 29 *CFR* 1910.147 and NFPA 70E state a preference for lockout of equipment over tagout.

☐　　☐　　4. Lockout/tagout training should be developed for a specific site and based on OSHA and consensus standards.

Answers: 1. (true), 2. (true), 3. (true), 4. (true)

Notes

1. Extract from OSHA Inspection Instruction STD 1.73.
2. This account is based on an actual incident. The names, including the name of the facility, have all been changed to protect those involved. Any similarity to actual names or facilities is strictly coincidental.

Chapter 10

The Electrical Safety Program

Definition

A safety program is an organized effort to reduce injuries. Any electrical safety program should be a subset of an overall site safety program. Processes and methods (e.g., inspections, audits, and feedback) that work in an overall safety program will work in the electrical safety program as well. In fact, these processes are required. However, electrical hazards, as well as the way people are exposed to them, are unique.

In most cases, mechanical hazards are observable provided an attempt to recognize them is made. A hole in the ground or a broken ladder can be seen if a person looks at them. Unstable objects normally are visible.

However, electrical hazards are not directly visible. The only way to "see" an electrical hazard is to observe signs and signals that indicate its presence. In general, people are exposed to an electrical hazard without their knowledge. The electrical safety program must contain elements that are designed to deal with this unique problem. Safety professionals are good at designing and operating an overall safety program. Electrical (and other discipline) technical professionals generally are required to support the safety professional to protect against electrical hazards. Neither safety professionals nor electrical technical professionals can be singly effective. Collaboration between and among disciplines is required for the program to be effective.

Training

Because electrical hazards are unique, training programs must enable people to "see" electrical hazards. The training also must provide information about how to avoid exposure to the hazard once it is recognized. The training must enable people to select and use personal protective equipment effectively. (For a detailed discussion of electrical safety training, see Chapter 5, "Strategies for Preventing Injury.")

Electrical work normally involves exposure to many other hazards. Electrical safety training, then, must be conducted in addition to any other safety training.

Up-to-Date Drawings

A strategy followed by most managers and production supervisors is to "keep the plant running and keep product moving out the door." Following that strategy, most managers and supervisors do not embrace the strategy of revising drawings to keep them up-to-date because that task costs money. (See Figure 10–1.)

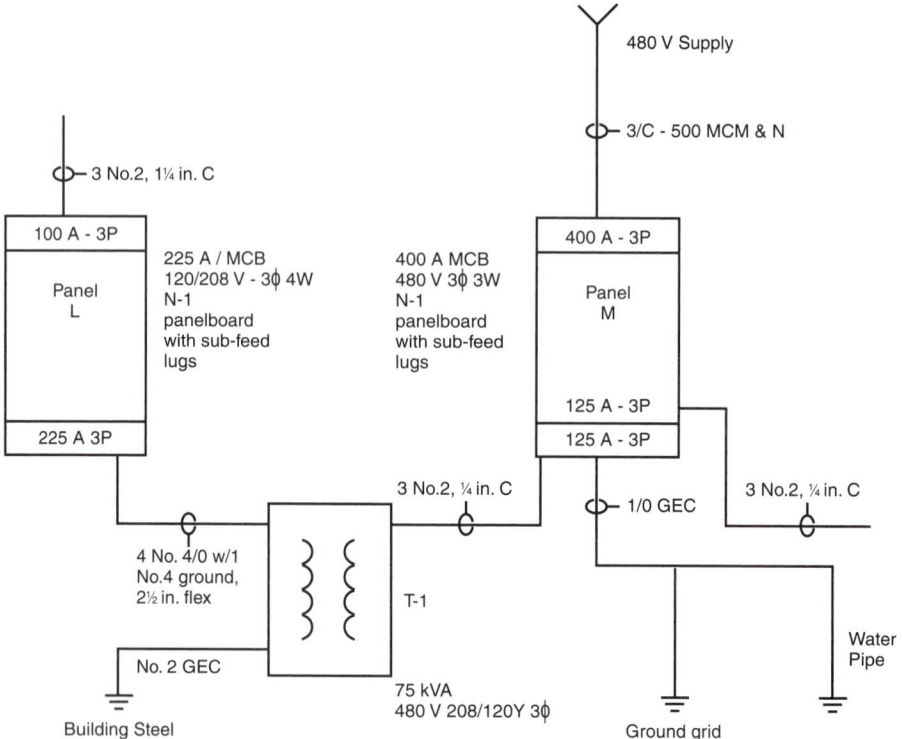

Figure 10–1. Typical Single-Line Drawing.
Source: Sargent, Jeffery S. and Noel Williams. *NFPA Electrical Inspection Manual with Checklists.* (Quincy, MA: National Fire Protection Association, 1999).

It is possible that the sheer number of drawings usually required for an electrical design elicit an expectation that all electrical drawings should be updated as changes occur over time. In reality, very few electrical drawings must be maintained for personal safety enhancement. Although a sound business investment, up-to-date schematic and wiring diagrams are not necessary for personal safety, but they enable troubleshooting to proceed expeditiously. *Electrical single-line diagrams are absolutely necessary.* They must be maintained in a current condition so that the source of energy is easily identifiable. A drawing maintenance system for single-line drawings should be in place and rigidly followed. The up-to-date drawings should be readily available for reference in order to reap the benefit.

Electrical drawing maintenance is not required by OSHA. However, NFPA 70E, *Standard for Electrical Safety Requirements for Employee Workplaces,* 2000 edition, does have a requirement for up-to-date single-line drawings. In industry, it is common to make changes in both the control scheme and the power-supply circuit during the functional life of a process. In fact, significant change can occur without making any change in the location of the disconnecting means. For example, the overcurrent protection scheme could change as product development advances.

The OSHA standards define a requirement for labels to identify the location of disconnecting means for lockout purposes. NFPA 70E also has a similar requirement. One of the first questions that must be answered to troubleshoot or maintain an item is: Where is the source of electricity? (See Figure 10–2.) Labels have a way of disappearing and are often incorrect. Doors and covers, with the label attached, may mistakenly be interchanged or misread. Alternate or temporary sources sometimes exist as well. Referring to an up-to-date single-line diagram drawing tells the worker at a glance where the disconnecting means is located. The single-line diagram also shows how all alternate sources may be isolated. The up-to-date single-line diagram shows temporary circuits, as well. Generally, three-line diagrams are much too complex and should be avoided.

The single-line diagram used for safety purposes must be simple, direct, and show all the pertinent information. The diagrams should show the following information:

- All distribution equipment and circuits
- Disconnecting means for each circuit

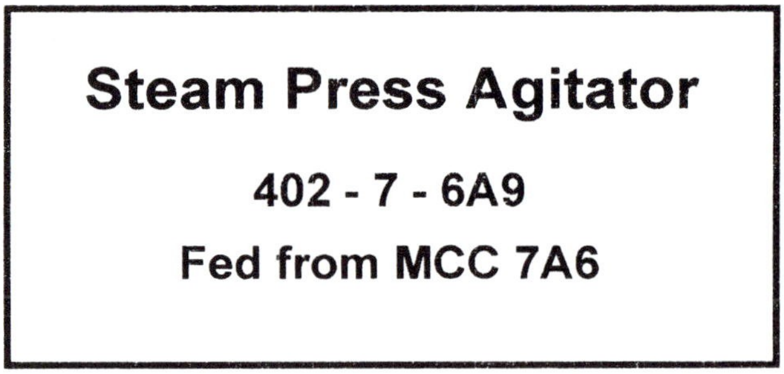

```
┌─────────────────────────────────────┐
│                                     │
│        Steam Press Agitator         │
│                                     │
│            402 - 7 - 6A9            │
│                                     │
│          Fed from MCC 7A6          │
│                                     │
└─────────────────────────────────────┘
```

Figure 10–2. Disconnecting Means Location Label.

- Utilization equipment
- Identification of equipment and circuits
- Wire size
- Overcurrent devices, including size and type
- Arc-flash boundary dimension (varies by location)
- Transformer size and impedance
- Circuit-grounding scheme
- Temporary circuits
- Plant, area, and process covered by the diagram

Procedures

What Should Be Included?

Although essentially undefined, procedures are a suggested requirement in many sections of the OSHA standards. NFPA 70E also defines a requirement for "procedures," without directly defining the word. Therefore, a standard dictionary becomes the source document for definition. According to Webster's unabridged dictionary, a procedure is "the act, method or manner of proceeding in some process or course of action." No connotation is made that a procedure is a written document. However, both OSHA and NFPA 70E require some procedures to be written.

The primary benefit of a written procedure is that it serves as a common communication tool. A methodology that is written provides the exact same set of words for people to develop their own understanding. It also overcomes a common propensity to forget steps or get them out of order. The written procedure enables an organization to improve the methodology as experience is gained by using the procedure.

Procedures have little value if they are not used and enhanced by experience. Otherwise, a procedure can quickly become outdated and, therefore, inaccurate. The existence of an inaccurate procedure is of negative value. Hazard exposure might be increased as a result of the inaccurate procedure. An employer's legal position is greatly undermined if procedures are out of date. Any OSHA fine likely will be cited as a willful violation if procedures are known to be out of date. If a procedure exists, it must be maintained and used.

Employers are required to provide procedures necessary to protect employees from injury. In OSHA terms, the employer must provide a safe workplace.[1] In NFPA 70E terms,[2] the employer must provide procedures and controls that are necessary to prevent injury to his or her employees. For a procedure to be effective, employees must follow the methodology and steps defined in the procedure.

Preventing injury requires a collaborative effort between employers and employees. Nowhere is the need for collaboration greater than in procedures. Employers are likely to know about energy and materials used at a plant or site. Employees are likely to know how to accomplish a task. Employers and employees are both likely to understand how to repair and maintain equipment and services. Suppliers and manufacturers also have specialized knowledge about electrical equipment. The best chance of preventing incidents and injuries is by integrating all this knowledge into a single written document.

Procedures should be audited on a regular schedule. That schedule should be more frequent when the procedures are new. The audit interval should not exceed one year. The audit should be designed to determine if the procedure does the following:

- Still addresses exposure to all hazards and how to avoid the exposure
- Is being correctly applied
- Has any shortcomings

A record of the audit should be established and maintained. The audit could identify a need for further training or a need to modify the procedure or the equipment. Any deficiency identified in the audit should be addressed in a logical way.

A procedure should provide all information needed to avoid inadvertent release of electrical energy. The procedural steps should not be followed blindly. Instead, the procedure should be applied after careful consideration. Any standard procedure step that is changed or not followed should be readily explainable. A procedure should, at least, consider the following ideas for content:

- Purpose of task
- Qualifications and number of employees
- Hazardous nature and extent of task
- Sequential listing of discrete steps to be taken
- Limits of approach
- Safe work practices to be utilized
- Personal protective equipment needed
- Insulating and special tools needed
- Special precautionary techniques
- Electrical diagram reference and location
- Equipment details needed
- Sketches/pictures of unique features
- Any reference information

Procedures as Part of the Overall Program

The electrical safety program should identify all electrical safety procedures that apply to the site or employer. Some procedures are clearly necessary. The number of OSHA-mandated procedures varies by the activity under way. At least two procedures are clearly required in 29 *CFR* 1910, Subpart S, of the OSHA standards: lockout/tagout and working on or near energized electrical services. Other sections of the OSHA standards identify other required procedures.

Several other procedures are probably necessary for an effective electrical safety program. Depending on the type of business, an employer might need to have a standard work procedure for inspecting small tools or for working in confined spaces. An effective electrical safety program will include a procedure covering all electrical work tasks that are electrically hazardous. The procedure should be written through a collaborative effort between the employer and employees.

> **NOTE:** Sample procedures are provided in Appendix B. These procedures are best used only as a starting point for site-specific procedures. The samples must be adjusted as necessary to address the manner and degree of potential exposure to hazards. The inclusion of the sample procedures is not intended to imply that the samples are the only procedures needed by a site. Additional procedures must be written to cover site-specific tasks.

Records

The electrical safety program must contain an element of records. OSHA requires some records to be kept and maintained. Some national consensus standards also identify the need for records. In Subpart S of 29 *CFR* 1910, OSHA requires that training records be maintained. Records of audits also should be maintained. The overall safety program should identify needed records and recording processes. In other words, although most of the training records will be on electrical training, they should be a part of the general program.

Program Content

Scope

Identifying the objective of the program and any intended boundary is very important. In reality, the scope of the electrical safety program is probably a subset of a larger safety effort that involves all disciplines. The scope statement is also important, unless the electrical safety program is merged into a larger program. Experience has shown that electrical hazards are no more prevalent than other types of hazards, but the consequence of a release of electrical energy is likely to be greater. The chance of a fatality or permanent disfigurement seems to be greater when the energy source is electrical. The electrical safety program should have a separate identity if it is part of an overall program. The scope statement should clarify any boundary and/or interaction.

Philosophy

The written electrical safety program should clearly state the intended philosophy for the organization. A philosophy tends to suggest basic corporate beliefs and support by management. It may also contain an element of strategy about how the philosophy may be executed. The following "fill in the blank" philosophy should be considered:

All injuries are preventable. Sound safety practices are a condition of employment and of continued employment at or on this site.

Responsibility and Expectation

The electrical safety program should clarify responsibilities and expectations, if any exist. The following statement of assigned responsibility and expectation should be considered:

Each person is responsible for avoiding exposure to a safety hazard and ensuring that all unsafe conditions are corrected. Each person is also responsible for identifying unsafe acts and for correcting such acts to the limit of his or her knowledge or ability. Each person is responsible for his or her own safety and for the safety of others. Each employee is expected to correct or report unsafe conditions or acts that are observed. Each person's attitude is reflected in his or her behavior. Each person is expected to know, understand, and use applicable safety procedures as tools to guide all tasks.

(See the sample electrical safety program in Appendix A.)

Controls

The written program should identify any controls that will be applied within the electrical safety program. Controls define how employees on the site are expected to think about situations. The following controls should be considered:

- No work shall be performed where exposure to hazards associated with electrical energy exists until an attempt is first made to shut down the source of energy.
- Every electrical conductor or circuit part is considered energized until proven otherwise.

- No barehanded contact shall be made with exposed energized conductors or circuit parts more than 50 volts.
- Deenergizing an electrical conductor or circuit part and making it safe to work on is, in itself, considered a potentially hazardous task.
- Each employer will provide procedures, and each employee will apply them to accomplish each task.
- Employees will be qualified for the task to which they are assigned.
- A hazard/risk analysis will be performed for each task involving any approach to energized conductors and circuit parts.
- The overall safety environment will be considered when assigning personnel to tasks.
- Each week will begin with a safety discussion.
- Each job line-up will include a discussion of hazards and procedures appropriate for the tasks involved in the job.

Training

The electrical safety program should identify all training that the employer provides to employees. (See Chapter 5, "Strategies for Preventing Injury," for a discussion of training.) Procedures that are expected to be implemented should serve one central element of the site's training program. As a procedure is discussed in the training program, the document becomes increasingly valid. Discussion of the requirements will eliminate any "soft spots" in the procedure content. The discussion also enables employees to feel included in the process, resulting in increased compliance with the requirements. Procedure training is an important element of discussion and communication between the employer and the employee as well as among employees.

Auditing

Auditing is a feedback mechanism that provides critical information to employers and employees as well. Audits are required by some national consensus standards and by OSHA standards. The basic idea is that only by auditing procedure implementation, work practices, and site conditions can an employer make adjustments, as necessary, to achieve the objective.

Some standards have a requirement covering auditing. The electrical safety program should, at the very least, include these requirements. Chapter 5 of NFPA 70E and OSHA lockout/tagout standards indicate that auditing at least one application of the lockout/tagout procedure is necessary. This audit (and any other) should attempt to identify "soft spots" in the process being audited. For example, the lockout/tagout audit should be designed to identify difficulties in either the procedure content or the employees' understanding of the procedure. Of course, if the audit identifies a weakness, it must be corrected.

Policies—Standards, Documentation, Excavation, and Emergencies

The electrical safety program should include any general policies that are either generated or endorsed by the employer. Every person working on the site or within the organization should implement program policies. Each employer should define and publish a policy covering standards, documentation, excavation, and emergencies.

Any standard that is embraced and implemented should be listed and named in the electrical safety program. For example, if an employer embraces OSHA Subpart S as one basis for the company's program, a policy statement should be written in the electrical safety program definition. The same concept holds true for all other standards.

If a policy exists related to drawings and other documentation, the policy should be included in the definition of the overall electrical safety program. Employees then have a clear understanding of expectations.

A policy statement should address excavations, because many services are buried in the ground. According to Murphy's Law (that says if something can go wrong, it will), if an excavation is made near an underground electrical service, the service will be interrupted. Applying a general policy avoids some of these incidents.

Sample Program

A sample electrical safety program is included in Appendix A. This program is not necessarily complete and may need to be adjusted before use. It is provided as a guide to fully develop an employer's programs.

 A Closer Look

Haines Electric[3] is a small employer in the Columbus area. Brothers Johnny and Jack Haines own the company as equal partners. Johnny is the administrator for the company, and Jack runs the field operation. Haines Electric had been profitable since it was founded 25 years ago, but, in recent years, profitability was diminishing. In fact, in the past year, Haines Electric had lost money for the first time in its history.

The workload had diminished because Haines Electric was being outbid on many jobs. Competitors were receiving many of the contracts that had once gone to Haines Electric. As the administrator, Johnny began searching the books in an attempt to explain why the company's bids were becoming less competitive.

Johnny's search didn't take long. As he looked at all the components that make up the cost segment of Haines' bid, it was obvious that next to the cost of labor, insurance constituted the largest cost. In prior reviews, Johnny had considered insurance as simply a fixed cost of doing business in the Columbus area.

Both Johnny and Jack treated their employees as friends. They were kind, considerate, and generous. It now seemed that the brothers would have to make some hard choices because the company simply could not afford to continue to lose money. The Haines brothers would have to make some changes or Haines Electric might have to go out of business.

Out of concern for the economic well-being of their employees, the Haines brothers asked all supervisors to gather in the shop one morning so that they could relay the poor business picture. Most of the supervisors did not understand how the situation could be so bad, when clients were so happy with their work. Johnny explained that the major problem was the high cost of insurance. One supervisor suggested a way to reduce insurance costs would be to reduce injuries and that would then reduce expenditures to the insurance company. In turn, reduced insurance expenditures would translate into reduced insurance rates. The idea made sense to both Johnny and Jack.

Over the years, employees at Haines Electric had experienced safety incidents and injuries from time to time. In the early days, Haines Electric employees were highly talented and trained. In fact, most of the supervisors were electricians when the company began. However, the type of work the company was doing had changed through the years. At first, the company concentrated solely on new construction, and their employees were particularly talented in construction activities. In recent years, however, manufacturers were trying to get more production from equipment and facilities already installed. As a result, the type of work shifted from new construction to improvements in existing facilities.

Employees were exposed to safety hazards in many different ways, as compared to the simpler hazards of the early days. The work environment had changed drastically over the years. Hazards do not change with a shift in work environment; however, in most instances, the degree of a hazard is likely to be different, perhaps greater and perhaps less. Not only is it likely that the degree of hazard will change but the degree of exposure will change as well.

In the supervisors' meeting, the brothers learned that Haines Electric effectively had no electrical safety program. A supervisor suggested the brothers approach one of their clients to ask for help. The selected client had an effective electrical safety program. In fact, the Haines Electric supervisor had been following the electrical safety program as a contractual requirement. The client suggested by the supervisor was happy to share knowledge and experience with the Haines brothers.

Johnny and Jack asked the supervisor to spend a few days in their office to help shape a Haines Electric safety program. It would emphasize the brothers' concern for their employees' well-being. It would include a policy that clearly stated that unnecessary exposure to electrical hazards is unacceptable. Their employees had to understand that decreased exposure to hazards would result in fewer injuries. The program would include training that emphasized personal responsibility and accountability, as well as employee involvement in generating and managing Haines' electrical safety program. The electrical safety program was only slightly modified from the same program implemented by the manufacturing location that shared the program initially.

Jack was instrumental in selling the program to Haines' employees. He mounted an offensive to make certain that all the program elements were implemented, including deenergization of equipment before beginning work and a special "test before touch" policy.

Safety incidents began to drop, even incidents not associated with electrical energy. Injuries almost disappeared. Insurance rates decreased significantly. Clients even began to award contracts to Haines based on the low number of incidents and injuries instead of the contract price. Haines Electric returned to a profitable position. In fact, the workload increased to the point that the Haines brothers had another strategic decision to make: How large should they allow the company to become?

Test Your Thinking

1. The only way to see an electrical hazard is
   a. By wearing a switchman's hood
   b. By observing signs and signals indicating its presence
   c. By wearing 3-D glasses
   d. None of the above
2. Up-to-date schematic and wiring diagrams
   a. Are a sound business investment
   b. Enable troubleshooting to proceed expeditiously
   c. Are critical for personal safety
   d. Both (a) and (b)
3. An effective electrical safety program
   a. Contains a procedure covering all electrically hazardous work tasks
   b. Contains a lockout/tagout procedure
   c. Outlines effective safety training
   d. All of the above
4. Auditing the safety program provides the following benefits:
   a. Identifies weaknesses in employee understanding
   b. Identifies weaknesses in procedures
   c. Identifies weaknesses in the line organization
   d. Both (a) and (b)
   e. All of the above

Notes

1. *The OSH Act, Public Law 91–596.* Section 5(a)—"Duties."
2. NFPA 70E, *Standard for Electrical Safety Requirements for Employee Workplaces,* 2000 edition, Part II, Chapter 1, Section 1-3—"Responsibility."
3. This account is based on an actual incident. The names, including the name of the facility, have all been changed to protect those involved. Any similarity to actual names or facilities is strictly coincidental.

Answers: 1. (b), 2. (d), 3. (d), 4. (d)

■ Chapter 11

Human Behaviors

Human behavior is the manner in which people conduct themselves. Behavior can be influenced by many things, not the least of which are human nature, education, experience, and example. Human nature can be defined as the fundamental dispositions and traits of people. Learning through education, experience, and example begins the day a person is born. All people learn through the education they receive and the experience they gain in every aspect of life. Parents or caregivers have a profound impact on children's education and behavior in the early years of their lives. In addition to formal education, people also learn by example. Through a mentoring process, people learn from counselors, tutors, or coaches. The apprentice program is one good example of a mentoring program.

In electrical safety training, behavior can be influenced in all of these ways. No one wants to be the victim of an electrical accident. However, the way individuals behave most certainly can affect not only whether or not they are exposed to electrical hazards but also their potential for accident and injury.

Implications of Behavior

One characteristic of human nature is to observe how another person behaves in common situations and then to make judgments about the person. When the person being observed is in a position of authority, people tend to accept the observed behavior as desirable. Making a value judgment of another person is almost always based on observations of the person's behavior.

A second characteristic of human nature is a tendency to emulate observed behaviors exhibited by a trusted coworker, colleague, friend, or acquaintance. When the observed behaviors are positive, emulation is a good thing. When the observed behaviors are less positive, however, the observed behavior is a negative influence.

Individual values influence a person's behavior. Each person in the workplace has a value system that is based on a value set that was taught to him or her in childhood. Sometimes, those values change as the person grows into adulthood and is able to think and rationalize. In the workplace, the natural tendency to emulate observed behaviors influences how, or if, individual values are guiding the observer's behavior.

In the workplace, when a supervisor or manager's conversations and queries with employees completely omit references to electrical safety issues, the employee recognizes that the supervisor or manager has low value for electrical safety. If the supervisor or manager never talked about hazards, a common observation would be that he or she is not concerned about employees' exposure to hazards. If a supervisor or manager tends to overlook safety violations, an employee might conclude that the supervisor or manager has little value for safety.

Of course, external observations can be inaccurate. A person's mind can easily be consumed by the normal pressures of operating a process, production quantity, time needed to repair a failed component, or the increasing unit-price cost for pulling cable. He or she might consider safety issues to be very important. Exhibited behavior will betray the real values almost every time.

The Workplace Environment

A work environment is made up of many different elements, including the people who work in the area. Groups of people tend to exhibit characteristics that might be similar to those of an individual. Assembled groups of people have a group personality and a willingness or unwillingness to take certain actions, regardless of whether the group is in a conference room or in a common area in a work site. The group leader's behavior has a significant bearing on the behavior of the group. The behavior of the other group members also influences group behavior, but not as much as the leader's. Sometimes, the group leader willingly relinquishes his or her leadership role to another member of the group. On occasion, a member of the group wrests the leadership role from the appointed leader. Most groups have members who are followers. They tend to be quiet and follow the lead of anyone willing to take it. The integration of all behaviors of group members results in an overall group behavior pattern.

Two characteristics that influence electrical safety in a workplace environment are the tolerance of the group for individuals' risk-taking behavior and the condition and arrangement of the physical area.

Tolerance for Risk Taking

From an electrical safety point of view, the group behavior tends to influence the degree of willingness of each member of the group to take a risk. When equipment doors are often left ajar or less than fully latched, it seems apparent that the group behavior is less than desirable. By their action or inaction, workers exhibit behaviors that are generally acceptable to the overall group. If generally accepted group behaviors include an intent to prevent injuries or incidents, the overall group incident or injury ratio will be low. However, if acceptable group behavior includes a willingness to take unnecessary risks, the incident or injury ratio will be greater.

Condition and Arrangement of the Physical Area

A second characteristic of the work environment is the condition and arrangement of the physical area. A work area that is highly congested will result in decreased ability of the people in the work area to recognize subtle shifts in the area. Some of those subtle shifts could be associated with broken grounding conductors, broken or missing connections, open doors, or corrosion. If people are not aware of changes in the work environment, they have no chance to avoid hazards associated with the change.

The behavior of people is also influenced by the physical arrangement of the workplace. A very common behavior is for people to congregate around desks and storage locations. Coffeepots are frequently installed in an electrical control room. By not reacting to this customary behavior, people are unnecessarily exposed to several potential electrical hazards that may exist in the control room. An organization should recognize and take action, as necessary, to circumvent such behavior. Desks and coffeepots can be placed at another location that has fewer potential hazards.

Secondary distribution equipment, such as motor control centers, is sometimes placed in an operating area. Because this equipment usually has a very high available short-circuit current, anyone near the equipment is potentially exposed to electrical arc blast and arc flash. Because the space in front of this equipment has no other equipment, a common behavior is to use this space for temporary storage. An organization should be aware of the tendency of such behavior and place this type of equipment in a more protected area.

Motivation and Behavior

To influence behavior, a basic understanding of motivation is necessary. Several theories have been promoted to explain or understand motivation. One such theory is Abraham Maslow's hierarchy of needs.[1] This theory suggests that a person has needs that he or she is driven to satisfy, beginning with basic physical needs. (See Figure 11–1.) The needs are sequential, and each must be satisfied individually before the next need becomes a motivating factor. The ultimate need is self-actualization, or self-fulfillment.

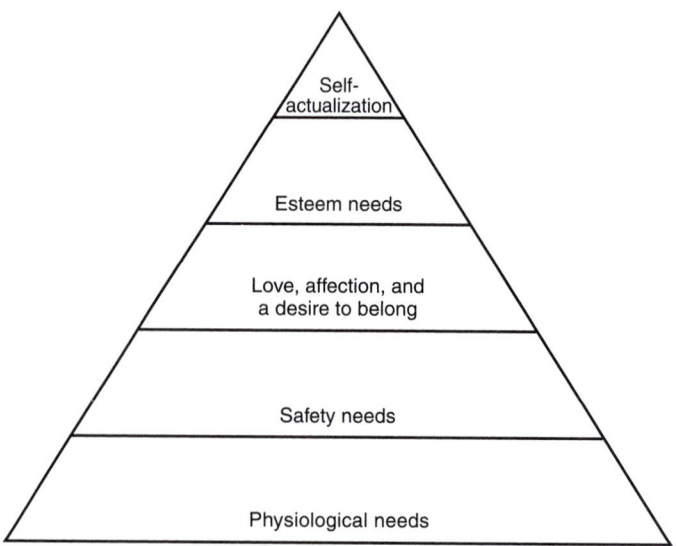

Figure 11–1. Maslow's Hierarchy of Needs.
Source: From Maslow, A. H. *Motivation and Personality* (2nd ed.). New York: Harper & Row, 1970.

- Physical needs—These needs include air, water, food, shelter, and physical comfort (e.g., warmth).
- Safety needs—These needs include freedom from an expectation of danger. They diminish as a person ages.
- Love, affection, and a desire to belong—These needs include feeling like a part of a group and giving and receiving love and affection.
- Esteem needs—People need to feel satisfied and have a high level of respect from themselves and from others.
- Self-actualization—In this state of self-achievement, a person feels that he or she has accomplished his or her "reason for being."

Behaviors tend to be aligned with actions that a person feels will achieve a greater sense of fulfilling one of these needs or progressing to the next level of need. Reinforcing a person's feeling of his or her position within this hierarchy provides positive motivation. Conversely, negative motivation tends to result in the person feeling as if he or she is being held back. In most cases, observing how a person behaves provides clues of where the person sees himself or herself in relation to Maslow's hierarchy.

Although some people still are striving to provide necessary needs—food, shelter, and physical comfort—in many cases, physical needs have been met. Money can provide for these needs and is a significant motivator. The motivator can influence behavior.

Whether or not a person feels freedom from danger depends on the person's understanding of how he or she may be exposed to danger. As electrical safety training builds greater understanding of hazards and exposure to hazards, a person tends to feel less safe. Electrical safety training can be a motivator that results in improved behavior. According to Maslow's theory, safety needs tend to diminish in importance as a person ages. Perhaps that is true; however, the observation could be the result of a diminished recognition of hazard exposure. Certainly, behavior is influenced by people's understanding of their safety needs.

A basic understanding of where a person sees himself or herself in relation to Maslow's hierarchy provides an indication of what the person believes to be the most value. Any action or suggestion that benefits the person also motivates him or her to behave differently.

In order to influence a person to modify his or her behavior, it is necessary to provide some motivation. Understanding what a person believes about himself or herself indicates what might provide the necessary motivation.

Unsafe Acts and Behavior

All injuries and incidents can be categorized as equipment failure, unsafe conditions, or unsafe acts. If all incidents categorized as equipment failure and unsafe conditions were lumped together and magically eliminated, only about one-third of the incidents and injuries would disappear. Unsafe acts are the basic cause of the remaining two-thirds. (See Figure 11–2.)

Unsafe acts are the result of poor behavior. Although many explanations are given for acting in an unsafe manner, most reasons are unacceptable. Training efforts and programs must include elements that are designed to influence behavior to increase awareness of these poor practices. The following list probably includes an explanation for the last incident you reviewed. Each explanation has a behavioral element.

- The person did not have the skill necessary to perform the task.
- The person did not recognize the risk.
- The person did not understand the consequences of his or her action.
- The person felt pressured by supervision or a customer to complete the task.
- The person was so familiar with the task that he or she was acting without thought.
- The person was distracted by a personal or family problem.
- The person was not familiar with some unique aspect of the equipment or circuit.
- The person was using an inaccurate reference drawing.
- The person did not recognize all the hazards associated with the task.
- The person was following an inaccurate procedure.
- The person was using faulty test or personal protective equipment.
- The person was not familiar with the job plan.
- The person did not understand the intended task.
- The person was distracted by another employee.
- The person did not use the right tool for the task.

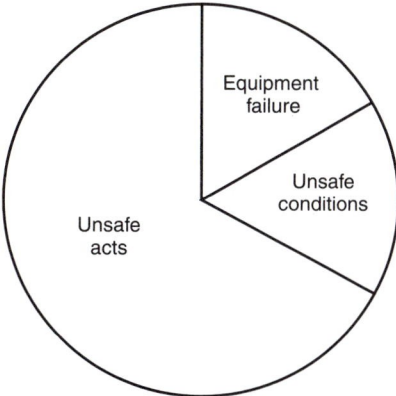

Figure 11–2. Illustration of Basic Injury Causes.

- The person had taken the same action many times in the past without incident.
- The person was acting in the way that his or her training has suggested.

Each incident and injury analysis should serve as a tool to identify the behavioral explanation for the event. Processes must be instituted that address each known poor practice. Most of the time, the poor practice will be behavior oriented. As suggested in the above list, an unsafe act should be considered systemic in nature, such as following a procedure that is out of date or relying on a drawing that is inaccurate. It is the unsafe act resulting from behavioral problems that can reduce incidents and injuries by at least two-thirds.

A Closer Look

It was about 3:00 P.M. on October 23. The sun was shining brightly when Alfred went to check on the work in the Poly building at the Gulf Coast MESP plant near Houston.[2] Alfred was a construction supervisor for SSC Construction. He was a valuable site representative for the company. He always paid particular attention to requirements that the customer included in the contract. Alfred knew that SSC was very serious about its lockout procedure. He also knew that the MESP lockout procedure required workers to remove all lockout devices at the end of the day, regardless of whether or not the work was complete.

The job in the Poly building was to replace a 45-kVA transformer with a new 75-kVA unit. No new wiring was required on the primary—both the wire and the disconnect size were large enough to supply the larger unit. New secondary wiring was already in place. The final task was to make the transformer connections in the transformer. The disconnect switch was the energy-isolating device, and it was locked at the beginning of each day. The MESP procedure included a requirement for a blue tag, representing transfer of ownership, to be installed on the disconnect switch to indicate permission for maintenance or contractors to install their lock.

The MESP procedure required each MESP employee to install a personal lock. Following OSHA's construction standards, Alfred had written an SSC procedure that permitted the supervisor to install locks to cover the construction employees.

Although Alfred listened carefully to MESP people, he frequently didn't adequately care for his own workers. On this day, quitting time for SSC electricians was 3:30 P.M., as the workday had started at 5:00 A.M. When Alfred reached the work area, he saw that the electrician was just finishing the primary connections. Alfred told the electrician to complete the primary connection and clean up. The new secondary conductors could be completed the next day. The position of the new larger secondary conductors made it very difficult to replace the front cover on the transformer. The electrician knew that MESP was adamant about lockout, so he did not worry about replacing the transformer cover.

As required by the MESP lockout procedure, Alfred proceeded to the panelboard and removed the SSC lock from the disconnect switch. At the time, Alfred thought it curious that the blue tag was missing, but the blue tag was not really his responsibility. He had no way of knowing that the blue tag had accidentally been torn from the switch earlier in the day.

During the next shift, lights were needed in the instrument control room of the Poly building. The second-shift electrician went to the panelboard to energize the lighting panel for the instrument control room. Seeing the switch turned off and seeing neither lockout locks nor blue tags, the shift electrician moved the switch handle to "on." However, the lights failed to come on. After a short investigation, the shift electrician reported back to the shift supervisor that the transformer was disconnected. The lights would not work that night. The work in the instrument control room would have to wait until another day. The shift electrician moved on to the work indicated on his next work ticket.

On Sunday, Matthew, the SSC electrician, entered the plant early in the morning. He was a new employee for SSC and wanted to do the best job he could. He knew what his work assignment would be that day, and he proceeded directly to the Poly building. Matthew did not question whether the circuit was locked out or not. He knew that MESP was adamant about lockout, so he prepared to begin work. Alfred arrived just before the work was started. Not really caring about the response, Alfred asked Matthew about the World Series game. Alfred was thinking about where to go next as Matthew replied that he had not seen the game.

Alfred was leaving the area when Matthew began to manhandle the new grounding conductor into position. The end of the new grounding conductor contacted the "A-phase" secondary tap on the transformer. The noise associated with the arc flash resounded throughout the Poly building. Matthew was wearing safety glasses, a hard hat, denim jeans, and a cotton shirt with long sleeves. His clothing did not ignite. He was treated at the local hospital for some burns on his hands and released. Again, an electrocution had been avoided only by luck.

 ## Test Your Thinking

True	False	
❑	❑	1. Employee behavior can be influenced by the value a supervisor exhibits for safety.
❑	❑	2. Both the tolerance of the group for individuals' risk-taking behavior and the condition and arrangement of the physical area can influence electrical safety in a workplace environment.
❑	❑	3. Leaving equipment doors ajar or less than fully latched is an example of a risk-taking behavior.
❑	❑	4. As electrical safety training builds greater understanding of hazards and exposure to hazards, a person tends to feel safer.
❑	❑	5. All injuries and incidents can be classified as equipment failure, unsafe conditions, and unsafe acts.

Answers: 1. (true), 2. (true), 3. (true), 4. (false), 5. (true)

■ Chapter 12

Cost of a Safe Workplace

Safety programs frequently are viewed as a cost to the employer without considering the expenditure as an investment. Determining the actual cost of a safety program versus the benefit is difficult. Although the return on investment for a safety program can be determined by the same analytical methods used to review any other investment, the elements included in the analysis are considerably different.

Little information about the cost of incidents and injuries is available to the public. Employers typically guard these records and treat them as confidential. Compiling this type of data is difficult, at best.

Normally, injuries affect an employer's profitability in a significantly negative way. The aftermath of a single injury can destroy a small employer. On a percentage basis, single injuries have less of an impact on large employers. However, single incidents can result in a decrease in stock value. Catastrophic incident results can have a catastrophic financial impact on a business.

Cost of an Incident

For this discussion, an incident is defined as an occurrence that resulted or could result in an injury or damage to equipment or the environment. An event in which equipment operates as designed is not considered an incident in this discussion.

The economics of an incident can be categorized using the following distinct results:

- Lost product
- Primary damage to equipment
- Shock, or weakening, of equipment integrity

- Damage to the environment that results in cleanup or litigation costs
- Decreased confidence of employees and supervision
- Damage to community relations

Lost Product

Because electricity is likely to be the source of energy that drives the process, product that would have been completed and sold will not be produced for a period of time. Any product that is produced probably will not meet expected quality standards or may, at best, be sold as reduced-quality product. In most instances, any product that was in production at the time of the incident will be "waste" product, resulting in disposal costs. Depending on the duration of the upset condition and the profit associated with the product, costs can vary from costs of recycling the waste product to expenditure of significant amounts of money for disposal.

Primary Damage to Equipment

Any time electrical energy is released unexpectedly, some damage occurs to equipment. Obviously, that damage varies from minimal to complete destruction. At the very least, the equipment must be inspected and cleaned before it is returned to operating condition. In some instances, the equipment must be completely replaced. As the degree of destruction increases, the length of time needed to repair the damaged equipment also increases. In some cases, spare equipment can be installed, and the production line can be returned to operating condition. However, even in these cases, significant dollars are required for equipment restoration.

An incident usually results in a high-stress situation where maintenance organizations, contractors, and employees are under pressure to rebuild the production facilities. In the high-stress situation, temporary repairs are undertaken. The temporary repairs usually are less than desirable and result in increased exposure to another failure or incident. A temporary replacement with a not-quite-right piece of equipment may result in amplifying additional exposure to a secondary incident or injury. Use of such a "temporary" replacement may also stretch into a longer time period than originally intended if the component seems to work adequately.

Shock, or Weakening, of Equipment Integrity

In most instances, a release of electrical energy involves an explosion of some degree. With the possible exception of explosion-proof equipment, electrical equipment is not manufactured to contain, or even withstand, an internal explosion. In addition to the maximum amount of pressure generated in an explosion, the time it takes to reach the maximum pressure has a bearing on the degree of enclosure damaged. In some cases, the equipment structure is destroyed. In other instances, the structure is deformed, resulting in decreasing the ability of the equipment to perform in the future.

Where the incident results in operation of an overcurrent device, damage is almost always done to the circuit. Where circuit breakers operate as a result of significant overcurrent, the circuit breaker's operation should again be verified before the equipment is returned to production. Where a fuse in one conductor opens, all phase fuses should be replaced, and the potentially damaged fuses should be destroyed.

After any accidental release of electrical energy, both the equipment and the electrical circuit should be thoroughly cleaned and inspected prior to their return to service. Both dollars and time are required to complete this important maintenance task.

Damage to the Environment That Results in Cleanup or Litigation Costs

Any time that waste product is produced, an environmental consequence may result. Waste product may be sent to a landfill or otherwise stored. Waste product may be recycled, but the additional processing will likely require additional energy, resulting in a different kind of environmental impact. An electrical incident can result in a significant release of materials into the air or ground water. Dealing with waste product requires that money be spent that otherwise would result in more profit for the company.

Environmental impact might also result in litigation. The litigation could result in an amicable resolution before significant court costs mount; however, avoiding the incident in the first place would have prevented such significant expenditures.

Decreased Confidence of Employees and Supervision

The confidence employees have in the integrity of the equipment tends to influence their behavior. As confidence wanes, behaviors tend to change, becoming slower and more deliberate—but not necessarily safer. Concern about exposure to a safety hazard increases. Maintenance costs increase. Response and repair times increase. Unfortunately, the increased concern may result in greater risk.

Damage to Community Relations

Factories and employers operate their businesses in a community only with the permission of the community, although that fact is not often realized or understood. A community either supports the existence of a manufacturing facility, or not. In most cases, a community is quite aware of incidents that occur within a facility. The type and number of incidents experienced by the plant influence whether the community continues to offer its permission or withdraws that permission. By controlling and reducing the number, type, and degree of incidents within the facility, a manufacturer or employer can influence how the community thinks about the existence of the plant. The cost of facility operation goes up as community confidence goes down.

Additional Economic Effects of an Incident

If an incident causes an injury to a person, the following negative economic affects can be associated with the incident:

- OSHA fines
- Criminal charges
- Litigation costs
- Increased insurance costs
- Retraining costs
- Costs related to investigating the cause of the injury

OSHA Fines

OSHA is authorized by congressional action to assess fines where an employer is cited, as in violation of specific requirements. Those fines can be as high as $70,000 per violation, per employee. Although multi-million-dollar fines are unusual, they are very possible. OSHA is chartered to enforce its requirements; therefore, the likelihood of action by OSHA increases with the severity of an incident, as follows:

- A visit from a field inspector is probable in the event of an incident that results in significant injury.
- A visit from a field inspector is likely if multiple injuries are involved.
- A visit from a field inspector is a certainty if the incident results in a fatality.

The size of a fine from OSHA depends on several factors, including past incident and injury experience at the facility. Avoiding the injury is the very best way to avoid any resulting OSHA fine.

Criminal Charges

A few obvious safety violations are well known to most employers. (An example is repeatedly locking doors that employees must use for exit in an emergency.) OSHA citations for this type of violation suggest that the employer has little or no concern for the well-being of the employees, and OSHA considers the employer to be engaging in "willful" violation. Although very rare, the OSHA citation can result in action in criminal court. Criminal charges can be assigned individually to the employer (the plant manager). Any fines resulting from such criminal charges may not be paid from employer funds. They must be paid from the personal funds of the employer.

Litigation Costs

Costs associated with litigation have no upper limit in the United States. Litigation resulting from company safety incidents is not only possible, but probable. Legal actions tend to occur over long periods of time. As the length of time increases, costs associated with the action also increase. In many cases, employers may find that "cutting their losses," so to speak, and reaching a settlement are less expensive than continuing the legal action for several years. Awards to an employee or the employee's survivors are almost always very high. All costs associated with litigation can be avoided by avoiding the incident and injury in the first place. An effective safety program significantly reduces this economic risk.

An employer's safety program assists either the complainant or the defendant. The program demonstrates intent. If it is effective, the defendant (company) has the advantage (demonstrating the company's good intent). If it is ineffective, the complainant has the advantage (demonstrating the company's indifference or poor intent). Any judgment will be influenced by the ability to demonstrate the intent of the company.

Increased Insurance Costs

Companies usually purchase insurance to help cover costs associated with workers' compensation, which is required by various legal jurisdictions. The insurance covers some portion of the weekly pay to workers who are injured on the job. The insurance company relies on experience to determine the cost structure of the coverage. As the number of injuries increases, claims paid by the insurance company also increase. To remain a viable business, the insurance company must increase premiums to the insured. The cost of the insurance is a significant factor for smaller companies. Some large corporations are self-insured. For self-insured corporations, workers' comp costs are even more visible.

Retraining Costs

If an injury occurs, the state of employee training should be reviewed. Some retraining is probably in order. The training program may be weak or ineffective. At the very least, some awareness retraining is needed. Although employees routinely should be provided with awareness training, any training immediately following an injury will be much more effective than at any other time.

Investigating the Cause of the Injury

If an incident involves an injury, determining the cause of the incident and ensuing injury as quickly as possible is important, especially if an investigation is initiated by an external agency. The investigation includes the following:

- An analysis of the site where the injury occurred
- Investigation of the state of the electrical installation and circuits that existed before the incident
- An analysis of the installation and schematic drawings that existed at the time of the incident
- A review of procedures and work practices that were in effect at the time of the incident
- Interviews of the people who were involved in the work at the time of the incident

All of these activities require time that otherwise would be consumed in executing tasks associated with operating the facility. Several people are involved in executing these tasks. A significant amount of energy and dollars will be consumed in determining the cause of the incident.

After the cause has been determined, changes to practices and procedures must be designed and implemented. Safety meetings and other training activities might be required. More expenditure is associated with these activities. During these activities, dollars are expended that would otherwise be used to produce a salable product.

Return on Investment

An effective electrical safety program requires an upfront investment, and program design costs money. Implementing an electrical safety program costs money, as does maintaining an effective program. People and resources are required.

People and resources are required to build a manufacturing plant and produce a product. However, these expenditures normally are considered to be an investment. A plant that produces a 50 percent annual return would be considered an excellent investment. Anyone would be happy with an investment that doubled in two years. Raise the stakes to double the invested money in one year, and everyone would participate.

According to the *Business Roundtable Report, Number A-3,* produced in 1982, the cost ratio of construction program to savings is 3:2. In other business segments, the reported cost-to-savings ratio is on the order of 4:1. One major corporation with an effective safety program suggests a cost-to-savings ratio of 5:1. These data suggest that money invested in an effective safety program is extremely effective.

Of course, there are limits to the return. When the point of diminishing return is reached, further investment tends to reduce the return. Designing and implementing an electrical safety program as described in NFPA 70E should provide maximum return on the investment. Program elements that exceed the program described in that document should be evaluated on the same cost/benefit basis as any other investment. However, the analysis should consider that the benefit of an electrical safety program is not always readily determined by the same analytical methods as equipment. An effective electrical safety program is a wise investment.

In many instances, safety programs are guided by requirements. When a company is implementing a safety program, it probably looks at requirements defined either by the insurance company or by requirements that are legally mandated. The company then establishes a program that meets, but does not exceed, the requirements. Unfortunately, this approach is typical. The cost of this type of program requires essentially the same people and resource demands that are required for a much more effective program. In order for an electrical safety program to provide a competitive advantage, a company must achieve a greater return from scarce resources. If the program is developed and implemented based on the need to keep people safe instead of the need to meet mandated requirements, the very small cost increase can produce the competitive advantage that is so desirable.

 # A Closer Look

Ralph Burdett worked as a production supervisor at Millrun Corporation's Cleveland plant, where plumbing fixtures were produced.[1] The Cleveland plant was being modified to produce a new fixture for kitchen sinks. In addition to the changeover for the new product, a new plating process was being installed.

Millrun employees were performing the changeover, but contractor employees were installing the new plating equipment. Due to potential health hazards associated with the plating process, the new equipment was separated from the general production facilities. The plating equipment was in a room adjacent to the general production facilities. A doorway was located between the production area and the new plating equipment.

Contractor employees had installed an extension cord through the open door. The cord was simply strung across the floor. Ralph considered the cord to be a tripping hazard, and he asked the contractor employees to get the cord off the floor, which they did. The new location for the cord was behind the door and above the top hinge. The contractor employees had reacted quickly to Ralph's request.

At break time, the contractor employees closed the door and left the area. Ralph noted that the door pinched the cord. With an "I simply don't believe that" approach, Ralph then asked James, another Millrun employee, to remove the cord from behind the door until the contractor employees returned. He decided he would have the contract employees install the cord through a pipe chase, as they should have done in the first place.

James opened the door and lifted the extension cord from behind the door. He did not notice the cut in the insulation on the cord. He probably would not have recognized the hazard even if he had seen the nick in the insulation. As his hand slid along the surface while lifting the cord, James contacted the exposed conductor. He was leaning against the grounded door frame. Electrical current flowed from his hand, through his chest, and exited through the wet shirt on his back. James made no sound. He died almost immediately.

James' family was advised by coworkers to contact a lawyer and initiate litigation, which they did. After all, the family would need money to get by now that James was gone. Both Millrun Corporation and the contractor were named in the suit. In the end, James' family was awarded $13 million in the wrongful death suit. Both Millrun and the contractor were assessed equally; although the contractor caused the hazard, Millrun was responsible for issuing the instruction that resulted in the fatality.

Test Your Thinking

1. Which of the following statements describes an electrical safety program?
   a. It is only a cost, with little economic benefit.
   b. It is a very sound business investment.
   c. It has no bearing on profitability.
   d. None of the above
2. An incident can have which of the following negative affects?
   a. Damage to equipment
   b. Damage to the environment
   c. Decreased confidence of employees
   d. All of the above
3. An injury can result in which of the following additional negative results?
   a. Criminal charges
   b. Litigation costs
   c. Increased insurance costs
   d. All the above
4. What is the possible cost/benefit ratio of an effective safety program?
   a. 3:2
   b. 4:0
   c. 5:5
   d. 3:2, 4:1, or 5:1

Note

1. This account is based on an actual incident. The names, including the name of the facility, have all been changed to protect those involved. Any similarity to actual names or facilities is strictly coincidental.

Answers: 1. (b), 2. (d), 3. (d), 4. (d)

■ Chapter 13

Safe Work Practices

"Test Before Touch"

Because the presence of an electrical hazard can only be seen by observing and reacting to indications and signals of its presence, one of the most important indications of a hazard is the signal suggested by a voltmeter. Electrical shock is the most widely recognized electrical hazard. Direct contact with an energized conductor frequently results in a fatality. Unless an electrical conductor is *known* to be deenergized, the conductor must not be touched without adequate insulating personal protective equipment (PPE). A voltage measurement must be taken.

The DuPont Company initiated a campaign called "test before touch" in the late 1980s and has shared the program with any organization that wants to use it. "Test before touch" can become the core of an extended campaign within a company. As conceived, however, "test before touch" is much more than a catchy slogan.

The concept is that every person tests every conductor, every time, for absence of voltage before they touch it. Critical to the concept are the phrases "every conductor" and "every time." Many recorded injuries involve a person testing a conductor for voltage, finding none, beginning work, then leaving for a cup of coffee or other reason. Upon returning to the work site and resuming work, the person receives a shock. Something changed while he or she was absent from the work site. With the "test before touch" concept in place, absence of voltage would again be verified upon the person's return to the work location.

DuPont's campaign features the logo in Figure 13–1. The company has willingly shared the campaign logo and concept with the general community through several IEEE conferences. The campaign has yielded a safety benefit for DuPont, a company well known for its ex-

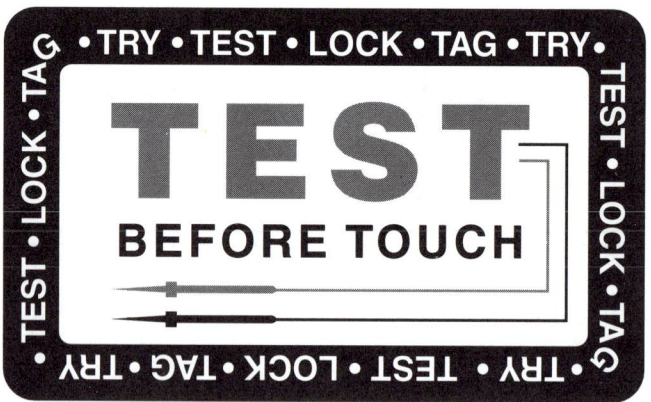

Figure 13–1. "Test Before Touch" Logo.
Source: DuPont "Test before Touch" Program.

cellent safety record. Any company could implement a similar campaign and reap a similar benefit.

Regardless of whether a "test before touch" campaign is initiated, testing for absence of voltage is an important work practice. No person should ever make contact with any electrical conductor until after he or she conducts a voltage test. If a person leaves the work site for any reason, he or she should conduct the tests a second or third time immediately before touching the conductor.

Blind Reaching

No instance exists where it is necessary for an employee to reach into electrical equipment without having direct line of sight. Many instances have occurred where a person reaches behind a cover or guard and contacts an energized conductor. In other instances, a person has great difficulty seeing the point he or she is trying to reach because of an obstruction. "Feeling for the terminal" could easily result in electrical shock. The ability to have direct line of sight with any point that will be or may be contacted is extremely important.

Implementing the "test before touch" work practice precludes blind reaching. Employers should formally prohibit blind reaching, and employees should faithfully embrace the prohibition. Both OSHA and NFPA 70E require that employees be instructed against blind reaching. Training programs should emphasize this point.

Awareness and Self-Discipline

It seems that "soft" concepts are more difficult to implement than more direct and observable concepts. Perhaps the "soft" concepts are more important. Safety-related work practices in both NFPA 70E, *Standard for Electrical Safety Requirements for Employee Workplaces,* 2000 edition, and OSHA, Subpart S, require that workers be instructed to be alert, however, "requiring" awareness is not possible.

It is possible, though, for supervisors to be familiar with people under their direction. If the supervisor has any indication that a worker is impaired in any way, the worker should not be assigned to work that involves exposure to electrical hazards. A physical impairment normally is easily observed. However, mental impairment is more likely to occur and harder to detect. An employee might have on his or her mind a spouse who is experiencing a medical problem or a child or other family member in some other kind of trouble. An employee might be experiencing a problem with drugs or alcohol. The supervisor should know enough about how a person normally acts and reacts to recognize any unusual behavior.

A person also can be directed to practice self-discipline. However, whether the person reacts to the instruction is not easily observable. As suggested in Chapter 10, "The Electrical Safety Program," an electrical safety program should identify principles and controls. Supervisors and other members of the line organization should behave in a way that exemplifies the principles and controls identified in the program. Training programs should encourage employees to think for themselves and instill the need for self-discipline. Employees should be encouraged to challenge the status quo. Tailgate discussions should cover ways to improve self-discipline.

Some tasks require work in the vicinity of energized conductors, where exposure to electrical hazards can occur. Employees assigned to these tasks must be continually aware of the hazard and should practice self-discipline. Working alone should not be permitted. At least two people should be assigned to tasks where exposure to electrical hazards is possible. If the degree of exposure is high, then another person should be assigned as an observer to warn the exposed person if he or she is approaching the energized conductor too closely.

Job Briefing

Normally, a workday begins with a supervisor assigning tasks for each member of the crew. The work assignment usually includes information such as reference drawings, tools, lockout points, test equipment, the nature of the problem, and the name of the person who reported the problem. If diagnostic steps are required, those steps are usually identified.

The task assignment must also include a discussion of all hazards associated with the work. The safety discussion should identify procedural safety requirements, PPE, how the worker may be exposed to hazards, and ways to minimize that exposure. The discussion should include emergency procedures (including the location of telephones and telephone numbers). (See Chapter 14, "Preparing for the Worst: Accident and Injury.")

In the utility industry, these discussions are called tailgate discussions. The discussion also might be called a safety task assignment (STA) or safety assignment first (SAF). Regardless of the name assigned to it, the discussion is critical. It is during this discussion that a supervisor has the last chance to note whether or not the worker seems mentally ready for the task. It is also during this discussion that the worker has the last opportunity to learn why the hazard has not been eliminated.

Illumination

Frequently, a task is assigned and the work begun without adequate lighting. In addition to mental awareness and physical ability, a person must be able to see the conductor or point where work is necessary. Task lighting must be supplied, as necessary, to enable the worker to see the entire area where a body part or tool might contact a live part.

Standards for lighting intensity probably do not exist other than recommendations by the Illuminating Engineers Society (IES). The intensity of the light does not have to be as strong as sunlight; however, the intensity must be greater than moonlight. The point is that the worker must have sufficient light to avoid unintentional contact with parts.

Some employees like to wear eyeglasses with tinted lenses. Some manufacturers' face shields and viewing windows also have tinted or reflective lenses. Wearing tinted lenses or face shields can increase the chance of inadvertently contacting unintended points. Increasing the chance of inadvertent contact increases the degree of exposure to electrical hazards. Both workers and employers are left with this conundrum: Does the increased protection afforded by the PPE overcome the increase in the degree of exposure from reduced visibility?

Conductive Articles on the Person

Both NFPA 70E and OSHA Subparts S and R prohibit uninsulated conductive articles from being worn when a person is working on or near open exposed energized electrical conductors. The objective, of course, is to avoid contact with the parts. The results of the contact would be shock and perhaps an arcing fault. In reality, all conductive jewelry should be removed.

People sometimes object to removing wedding rings. However, rings can be rendered nonconductive by the addition of insulating material. A worker could wear voltage-rated gloves, or the rings could be taped with electrical tape. In all likelihood, however, voltage-rated gloves should be worn because, by definition, the work is on or near live parts.

Eyeglass frames frequently are made of a conductive material. It is unlikely that a person's face would be close enough to contact a live part. It is possible, however, for the eyeglasses to fall from the face and onto a live part and initiate an arcing fault. To prevent this, eyeglasses should be restrained.

Some articles of clothing are either inherently conductive or made conductive by the addition of reflective material on the surface of otherwise nonconductive clothing. The addition of the conductive material usually is meant to improve the thermal characteristics of the clothing. In areas where a high temperature thermal source exists, a metallized surface added to the clothing substantially improves the thermal protective characteristics. However, the added conductive surface substantially increases degree of exposure to an arc flash in the vicinity of live parts. Conductive clothing must be avoided in an electrical environment.

Covers, Shields, and Barricades

Because electrical shock has been known to be a hazard for many years, electrical equipment manufacturers have attempted to provide a barrier on the line side of equipment to eliminate the possibility of accidental contact. Industrial purchasers of electrical equipment request the line-side barriers. The degree of protection afforded by the installed barrier varies from one manufacturer to another. In most instances, the barrier does prevent direct contact with the line side. However, the protective barrier also prevents direct contact with a voltmeter or tool. To gain access to the line side, electricians and technicians remove the barrier and sometimes do not replace it when their task is complete. If the barrier is not replaced, all possible protection is negated. Barriers that were installed by a manufacturer must be kept in place.

Electrical equipment is constructed so that isolating shields exist between discrete circuits. For example, individual starters and switches in a motor control center (MCC) or circuit breakers and switches in secondary switchgear are isolated from each other. Some of this shielding is required by the testing laboratory. Purchasers sometimes increase the requirements for isolating circuits and feeders. The objective is to avoid the possibility of a failure in one component from migrating to another. The integrity of these shields and barriers must remain as manufactured. Wiring that must be installed within the MCC or switchgear must not penetrate any installed shielding.

During the life of electrical equipment, installation of a new device probably will require that a hole be cut in the door or cover. Frequently, an existing device is no longer needed with the addition of the new device. A push button, pilot light, or meter might be removed. In these instances, formerly used holes are left open after the device is removed. After all, the equipment is in an enclosed room with limited access. However, any hole or penetration in the front of the equipment provides a ready path for gases and plasma to escape and increases exposure for any person standing nearby.

Adding a temporary barrier or shield in addition to those installed by the manufacturer might be possible. Selectively adding rubber blankets or other rubber products can also serve to minimize exposure. On occasion, clean dry wood can add a degree of protection from accidental contact with energized components. However, the addition of this or other temporary protection will probably involve a short-duration increase in hazard exposure. Although temporary shields and barriers can be effective, the temporary increase in exposure must be compared with the degree of later protection before the choice is made.

Covers and shields are intended to eliminate or minimize exposure of a worker to an electrical hazard. Sometimes, observers and other persons are also exposed to electrical hazards. For example, a supervisor attempting to answer a question or provide direction is frequently within a minimum approach boundary. A person nearby who is working on a different task might be within a minimum approach boundary. (See Table 13–1.)

The work area that contains electrical hazards should be within a barricade or similar physical warning mechanism. Before opening

Table 13–1
Approach Distances to Exposed Energized Electrical Conductors or Circuit Parts

(1)	(2)*	(3)*	(4)*	(5)*
	Limited Approach Boundary		Restricted Approach Boundary; Includes	
Nominal Voltage Range Phase-to-Phase	Exposed Movable Conductor	Exposed Fixed Circuit Part	Inadvertent Movement Adder	Prohibited Approach Boundary
0 to 50	Not specified	Not specified	Not specified	Not specified
51 to 300	10 ft. 0 in.	3 ft. 6 in.	Avoid contact	Avoid contact
301 to 750	10 ft. 0 in.	3 ft. 6 in.	1 ft. 0 in.	0 ft. 1 in.
751 to 15 kV	10 ft. 0 in.	5 ft. 0 in.	2 ft. 2 in.	0 ft. 7 in.
15.1 kV to 36 kV	10 ft. 0 in.	6 ft. 0 in.	2 ft. 7 in.	0 ft. 10 in.
36.1 kV to 46 kV	10 ft. 0 in.	8 ft. 0 in.	2 ft. 9 in.	1 ft. 5 in.
46.1 kV to 72.5 kV	10 ft. 0 in.	8 ft. 0 in.	3 ft. 3 in.	2 ft. 1 in.
72.6 kV to 121 kV	10 ft. 8 in.	8 ft. 0 in.	3 ft. 2 in.	2 ft. 8 in.
138 kV to 145 kV	11 ft. 0 in.	10 ft. 0 in.	3 ft. 7 in.	3 ft. 1 in.
161 kV to 169 kV	11 ft. 8 in.	11 ft. 8 in.	4 ft. 0 in.	3 ft. 6 in.
230 kV to 242 kV	13 ft. 0 in.	13 ft. 0 in.	5 ft. 3 in.	4 ft. 9 in.
345 kV to 362 kV	15 ft. 4 in.	15 ft. 4 in.	8 ft. 6 in.	8 ft. 0 in.
500 kV to 550 kV	19 ft. 0 in.	19 ft. 0 in.	11 ft. 3 in.	10 ft. 9 in.
765 kV to 800 kV	23 ft. 9 in.	23 ft. 9 in.	14 ft. 11 in.	14 ft. 5 in.

*All dimensions are distances of live part to employee.
Note: Flash protection boundary must be calculated in accordance with NFPA 70E, Section 2–1.3.3.2.
Source: NFPA 70E, *Standard for Electrical Safety Requirements for Employee Workplaces*. Quincy, MA: National Fire Protection Association, 2000.

doors or removing covers, the electrician or technician should erect the warning device. The barricade should include signs warning anyone approaching the work area to stay outside the approach boundary unless wearing necessary protective equipment.

Conductive Tools and Equipment

From an electrical shock point of view, a conductive tool held in a person's hand can be thought of simply as an extension of the hand. A person holding a ratchet handle that touches an energized conductor will experience the same result as if the conductor is touched with his or her hand. A shock is certain; electrocution is possible; an arc flash is likely. The hazards are all present. Conductive tools should not be used in the vicinity of live parts. They should not be used at all unless an electrically safe work condition exists.

Electricians and technicians often apply tape to hand tools. Additionally, manufactured comfort handles are on some tools, such as pliers and crimping tools. These practices provide no assurance that the user will be protected from electrical shock. Although electrical tape has good insulating qualities, it is not intended for this use. Although purchased comfort handles are plastic and have some insulating qualities, they are not intended for this use, either. Unless a tool is insulated, it will not protect the person from electrical shock.

ASTM F 1505[1] is a consensus standard that discusses insulated hand tools. The requirements of this standard are patterned after similar products in the European marketplace. Tools meeting ASTM F 1505 requirements are marked with the voltage rating. One issue related to this product is how to test the product after use to ensure that the insulating integrity remains. The tools are usually constructed of two-color materials. The inner layer of one color provides adequate insulation, and whereas the outer layer provides the wear surface. When the outer layer wears or is damaged to the point that the inner layer becomes visible, the integrity of the insulation is damaged. The tool should be discarded and destroyed.

Insulated hand tools must not be depended on as protection from shock. These products help avoid the initiation of an electrical incident, and they are valuable for this purpose. However, more conventional products, such as voltage-rated gloves, are much more reliable.

The best combination is to use insulated tools in conjunction with other PPE as determined by the hazard analysis.

Anticipating Failure

Sometimes electrical equipment provides a warning of impending failure. Hissing or crackling sounds usually indicate a significant leakage. Of course, the familiar 60-Hz hum is exempted. When these noises are detected within a circuit breaker or disconnect switch, a strong chance that a failure will occur in the near future is indicated. Simply moving the operating handle can cause the failure to occur.

The best approach is for the worker to follow this procedure:

- Post a warning for other persons.
- Remove the load from the circuit breaker or switch.
- Be sure that the door is closed and completely latched.
- Select the necessary PPE, such as flame-resistant clothing and voltage-rated gloves.
- Stand to the side of the compartment and move the operating handle to "off" before investigating.

An unusual odor in the vicinity of electrical equipment also can be a warning of impending failure. When a component is overheated, a unique odor may seem to be emanating from inside the equipment. The stronger and more pungent the odor, the greater the chance of equipment failure. The source of the odor should be identified, if possible. However, the equipment doors and covers should not be disturbed until either the source of the odor is found to be external to the equipment or the equipment is deenergized.

Temporary Grounds

For many years, temporarily grounding an electrical conductor has been accepted as the "final" check before contacting an electrical conductor that has been energized. "New" conductors that have never been energized normally are not grounded. Many electricians learn the hard way that medium-voltage electrical cables have capacitance and will regenerate a voltage even after they have been grounded out.

A period of time is required for the charge to drain from the long capacitor. Other electricians learn on the job that a long electrical conductor that lies or hangs parallel to another energized conductor can generate a voltage by transformer action. In still other instances, electricians find that voltage can appear on a conductor that previously has been tested for absence of voltage by an energized line falling on the "deenergized" line under repair. An atmospheric static discharge can also be the cause of a voltage reappearing on the "deenergized" conductor. In all of these instances, it is important to install grounds on a conductor before contact is made.

In industrial plants, temporary grounds are sometimes installed on the wrong conductors, resulting in large faults. Temporary grounds may have been constructed with welding clamps and installed on large conductors. These ground sets are not capable of tripping the overcurrent device. Grounding sets with large clamps have been installed in small spaces, and temporary grounds without clamps have been used to "drain" a charge from a medium-voltage cable, resulting in significant exposure to shock or blast. Circuits have been energized with "temporary" grounds still in place.

Temporary grounds are both a boon and a menace to electricians. It is important to understand the need for temporary grounds. It is also important to understand the need for high integrity of temporary grounds.

An electrically safe work condition cannot exist without analyzing the need for temporary grounds and installing them where necessary. Outside overhead lines can be reenergized by a number of possible methods. In every instance, temporary grounds should be installed where work is being performed on outside overhead lines.

Where shielded cables are installed underground, in cable tray, or in conduit, a temporary ground is in order, at least until any capacitive charge is relieved. In instances where the insulated cable ends are both contained in enclosures, the possibility of a voltage reappearing on the conductors is dramatically reduced. If the conductors are unshielded, the possibility for being surprised by unexpected voltage is almost eliminated, and grounds are usually not necessary. Even so, a test for absence of voltage is still an invaluable work practice.

For several years, training programs taught a practice called "chaining." As it was taught, the practice of "chaining" suggested that the last act before contacting an energized electrical conductor (especially with overhead construction) was to throw a log chain across the line. If the chain stayed on the line, it was deenergized. If the line was still energized, the chain would be thrown off in a large fireball. This, of course, was a terrible practice—the large fireball injured many people. However, the people involved reasoned that they had prevented shock and possible electrocution.

Electrical current flowing in an electrical conductor results not only in the movement of electrons but also in a magnetic force. As the current flow increases, the amount of force also increases. In the event of an electrical fault, the amount of current flow is very large, resulting in a huge force being applied to the grounding conductor. The integrity of the grounds must be sufficient to remain in place if the grounds are subjected to current flow. The grounds must remain in place for a longer period of time than that required for the overcurrent device to operate. Ground conductors (sets) should be constructed as defined in ASTM F 855.[2]

If a person is in contact with an electrical conductor when the conductor becomes reenergized, a large amount of current flows in the grounding conductor. Because the ground set is essentially installed in parallel with the remainder of the circuit, some significant current flows in the circuit. While current flows in the grounding conductor, a difference of potential is likely to exist along the conductor where the person is working and at another point along the path of the current flow. Any difference of potential between two points where the person is touching (or standing) results in exposure to either touch or step potential. Electrocution is possible even with ground sets of adequate integrity. The safe work practice is to install grounds in a way that establishes a zone of equipotential all around the worker.

In both an industrial and utility environment, a problem exists in making sure that any temporary ground that is installed is also removed. Injuries have occurred and equipment has been destroyed when a circuit is reenergized with a temporary ground still in place. Ground sets should be assigned a unique identifier, such as a code number. Each ground set should be assigned a storage location and position with the same identifier as the ground set. Workers should maintain a reference list that shows where each ground set is installed. In fact, the single-line diagram is a very good place to show each ground set and where it is installed. When the work is complete, the reference list should be reviewed and marked each time a ground set is removed.

Resetting Protective Devices

Electrical utilization equipment is required by code to have short-circuit protection. Motors and similar equipment generally have overload protection as well. These types of protection are intended for different failure modes. Short-circuit protection is intended to operate if an insulation breakdown occurs, whereas overload protection is intended to avoid an electrical failure that can result from overheating when a motor is mechanically overloaded.

Protective devices can operate due to a temporary condition, such as a conveyor that is temporarily jammed. When a motor or other electrical utilization equipment fails, the first maintenance action normally taken is to push the overload reset button. If the circuit is not restored, the second normal maintenance action is to replace fuses, where they exist, or reset the circuit breaker. Although these actions may be required relatively frequently, they involve some degree of exposure to electrical hazards.

Before a person pushes a reset push button, he or she should observe that all door and/or cover fasteners are in place and latched. Incidents are on record where a major fault occurred when the overload reset button was pushed. Resetting a tripped circuit breaker demands the same advance observations.

If a fuse change is necessary, of course, doors must be opened. Prior to opening the door, the person must be wearing the necessary PPE. On many occasions, a significant fault results from the action of opening a door. Springs that slip from their mount or break from stress fracture can easily initiate a fault.

Protective devices that operate a second time must not simply be replaced and put back into service. The repeat operation indicates that the problem condition is not temporary. The operator must diagnose and correct the problem before placing the circuit back into service.

Approach Boundaries

One of the best ways to prevent an injury is to maintain space between a person and the hazard. As the distance (space) from an electrical hazard and a person decreases, the degree of exposure increases. It is generally believed that as a person becomes more qualified, he or she has a greater chance of avoiding injury. Some consensus standards (and some people) identify only a single safe approach distance for all types of work. However, identifying a safe approach distance is not that simple. A single safe approach distance deals only with one hazard—shock—when multiple hazards and multiple exposures may exist. NFPA 70E and the authors tend to embrace a more conservative approach, which is to identify safe approach distances that depend on each hazard (shock or arc flash) and the qualification of the person.

The concept of boundaries is that any point A on an exposed, energized electrical conductor or circuit part serves as the center of an imaginary circle around the point. Imaginary circles at various distances surround point A. The distance to the imaginary circles then defines the location of approach boundaries, as illustrated in Figure 13–2.

The distance to the *limited approach boundary* is the point where any person, regardless of qualifications and ability, is subjected to a shock hazard. Any person outside this imaginary circle is considered safe from exposure to shock. Therefore, this boundary defines the closest

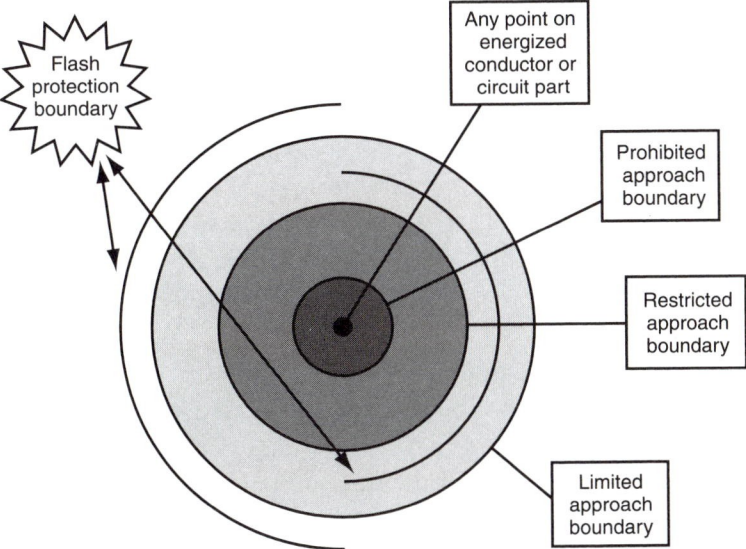

Figure 13–2. Approach Boundaries Concept.

safe approach for an unqualified person. All unqualified people must remain outside this boundary unless they are under the tutelage of a qualified person. A person who is undergoing training may be within the limited approach boundary, provided he or she is under the direct supervision of a qualified person. Table 13-1 on page 183 shows the limited approach boundary for various voltage ranges.

The distance to the *restricted approach boundary* is the point where some controls should be put into place for qualified persons (see Table 13–1). This is the point where any person, regardless of qualification, is sufficiently close for the degree of exposure to be great. A second person should be nearby. This degree of exposure should only be accepted after an authority in the line organization has denied a shutdown request.

Any person crossing the *restricted approach boundary* should have a plan of action that has been accepted by a person in authority. The plan should specify that the person conduct a hazard analysis, minimize all possible exposure by adding barriers, and maintain as much space between himself or herself as possible while executing the work.

The distance to the *prohibited approach boundary* is the air breakdown distance, plus a small dimension for a margin of safety. The distance to the prohibited approach boundary is very short. The meaning of the words in the term *prohibited approach boundary* have been carefully considered and selected. The term means just what is says. Any approach with hands or tools closer than this distance is prohibited. Only qualified persons who have fulfilled the requirements of all controls and who are wearing the necessary PPE may cross this boundary.

Any person crossing the *prohibited approach boundary* should already have executed a hazard analysis, have a plan covering what he or she is going to do, and have authority (usually from a high-level manager). Of course, the person must be qualified for the task being attempted.

The *flash-protection boundary* is defined as the distance from a potential arc source beyond which a person may be exposed to a flash-burn injury. Any person who crosses this boundary must be protected from exposure to arc flash. The flash protection worn by the employee must be selected based on the amount of energy that is available (see Chapter 2, "Electrical Hazards").

Body Position

In many instances, while attempting to execute electrically hazardous work, a worker does not consider that he or she might position his or her body in such a way that minimizes exposure to the electrical hazard. A person might attempt to overreach and increase the propensity to contact a live part. The person might kneel in front of a cubicle in order to improve access to devices located near the bottom of a compartment and instinctively move when a pebble or another foreign body on the floor happens to cause a sharp pain.

Before beginning the work, it is a good practice for a person to take a few minutes to think about how the body will be positioned while executing the work, and he or she can decrease his or her hazard exposure. It would be very unusual for a plan to detail body position, but the worker's thought processes should include body position.

 # A Closer Look

Charles Andrews was widely known as an electrical expert. More than that, he was internationally recognized as an electrical safety expert, and he was widely respected for his ability and knowledge. Charles was the person who was instrumental in defining requirements for isolation between circuits and compartments several years earlier. He was accepted as the electrical expert at S&R Corporation, a multinational company of over 150,000 employees.[3] S&R's chemical and petroleum products included everything from food, to packaging, to clothing.

Charles was the leader among a large group of engineering consultants for the corporation. He had earned the admiration of all his colleagues at S&R as he advanced through the ranks in the electrical discipline. He had served as an electrical superintendent on several construction projects, as well as the lead design engineer for several other projects. Charles had been maintenance supervisor at four plant locations. Since accepting the assignment in S&R's internal consulting organization, he had developed the maintenance program to the point where failures in the electrical distribution system were effectively nonexistent.

Charles shared his expertise with the broad community. He actively participated in both United States national and international standards. Experts in other segments of industry frequently sought his advice. In every way, he was an expert's expert.

On a Monday morning in May, Charles was again sharing his knowledge and expertise by leading a group of maintenance consultants in performing annual maintenance at the S&R plant site in Monterey, Mexico. The chemical plant used batch-type processing. Although a single batch process could be shut down for maintenance, this annual maintenance required a total shutdown of the facility, including the administration building.

Charles and his colleagues arrived on the plant site at 6:00 A.M., ready to go to work. The plan was to allow production to continue until 8:00 A.M., then gradually to take the plant down. Everyone not needed for the annual maintenance would have the day off. The electrical maintenance electricians were to be supplemented by contract electricians for the day. The electrical maintenance would be completed by midnight on Monday, and the plant would start production again on the midnight shift Tuesday morning.

Charles was always anxious to get started and frequently pushed the envelope. He was the expert, and no one questioned his judgment. One task that he always recommended was to check the tightness of cable terminations. Charles was with three colleagues looking at substation 12-6A and talking about what work would be needed that day in this particular substation.

The substation had a grounded wye transformer, so each conductor was 277 volts to ground and 480 volts between phase conductors. The substation was an outdoor unit. Stone had been spread on the ground to enhance drainage of rainwater from the area.

Charles left the group of colleagues and went around the back of the substation to open one rear compartment door. As was his frequent custom, he grasped one cable and gave it a tug, checking to see if the aluminum cable terminations had loosened by the normal heating and cooling cycles in use.

His colleagues in front of the substation heard no sound. However, when Charles did not return after several minutes, one of his colleagues went to look for him. Charles was lying on the ground, and one rear door of the substation was open. The colleague called to the others for help and began CPR immediately. The rescue squad was summoned, and they responded in just over three minutes. But Charles was dead before they arrived.

The autopsy revealed that Charles had a stone bruise on his right foot. No one knows for sure, but the investigation suggested that he instinctively moved when he stepped on a stone on the ground, which put pressure on the bruise. That movement apparently resulted in contact with a live part as he tugged on the cable. Charles had not considered that his body position had significantly increased his exposure to electrical shock.

Charles was never replaced in S&R Corporation. It is not possible to replace the experience and ability that Charles accumulated through his 45-year career.

 ## Test Your Thinking

True   False

☐   ☐   1. Unless an electrical conductor is known to be deenergized, the person must not touch the conductor without adequate insulating PPE, and he or she must take a voltage measurement.

☐   ☐   2. "Feeling for the terminal" is an accepted practice if a person does not have a direct line of sight to the point he or she wishes to reach.

☐   ☐   3. An electrically safe work condition cannot exist unless the person analyzes the need for temporary grounds and installs them where necessary.

☐   ☐   4. Wearing PPE to open a door and change a fuse is unnecessary. In addition, a protective device that operates or trips two or more times may simply be replaced and the circuit restored to service.

☐   ☐   5. Only qualified persons who have fulfilled the requirements of all controls and who are wearing the necessary PPE may cross the prohibited approach boundary.

Notes

1. ASTM F 1505, *Standard Specification for Insulated and Insulating Hand Tools.*
2. ASTM F 855, *Standard Specification for Temporary Grounding Systems to Be Used on De-energized Electric Power Lines and Equipment,* 1997.
3. This account is based on an actual incident. The names, including the name of the facility, have all been changed to protect those involved. Any similarity to actual names or facilities is strictly coincidental.

Answers: 1. (true), 2. (false), 3. (true), 4. (false), 5. (true)

Chapter 14

Preparing for the Worst: Accident and Injury

The key to an excellent safety record for any industrial site is to prevent accidents and injuries. A well-planned electrical safety program with procedures, education, and training is the foundation for a safe work environment. However, according to the U.S. Department of Labor, an average of more than 4,000 nondisabling and more than 3,600 disabling electrical contact work-related injuries are recorded annually in the United States. In addition, although no statistics are available, electrical experts believe that arc-flash burn injuries may number three times these recorded figures. As previously discussed, arc-flash injuries are often recorded only as "burns" and not as electrical incidents. (See Chapter 2, "Electrical Hazards," and Chapter 5, "Strategies for Preventing Injury.")

Electrical accidents and the complexities of the trauma caused to the human body, not only physically but also mentally, have not been well understood in the past. The extent of injury to the victim often is not immediately apparent. Some symptoms may be masked by the more readily apparent thermal effects of the injury. As more knowledge is gained about electrical trauma, strategies for effectively handling the emergency and ways to improve treatment of victims have emerged. In addition, research suggests ways that workplace supervisors and responders can help an accident victim's caregivers provide appropriate medical attention.

Putting a Plan in Place

Having a plan to follow in the event of an actual accident avoids panic and loss of valuable time. Sites should establish a training policy and plan to cover electrical accident rescue methods, approved rescue devices, and CPR training.

CPR and First-Aid Training

Advance training of personnel in CPR and first-aid techniques is absolutely essential for accident preparedness. CPR training and periodic retraining for personnel should be carefully planned and documented. Personnel who are CPR trained should be clearly designated, and trained persons should be readily available.

A well-stocked first-aid kit in a weatherproof container should be maintained at all times. The kit should be kept in a prominent, readily accessible location that is well known to all workers. Supplies in the first-aid kit should be approved by a consulting physician and maintained with individual, sealed packets for each type of item. Contents of the kit should be checked at least once a year to ensure that supplies are present and in good order.[1]

Locations of eye-wash stations and showers should be prominently posted so that victims' burns can be quickly cooled and flushed after an accident.

Transport for the Accident Victim

Plans should be in place for transporting accident victims to a physician's office or hospital as quickly as possible. The lives of electrical accident victims may be saved and their recovery enhanced if they can be transported quickly to a burn center or other medical facility that specializes in electrical trauma. Industrial sites should evaluate medical facilities in their areas and determine in advance where such victims should be taken and how they will be transported. Having an ongoing relationship with personnel at the medical facility is an excellent idea. Emergency telephone numbers and specific instructions should be conspicuously posted.

Effects of Electrical Accidents on Victims

In a workplace high-voltage accident, the victim's skin, ears, eyes, lungs, internal organs, and nervous, muscular, and skeletal systems can be affected not only by the direct effects of electrical current but also by the following:

- Radiative heat from an electrical arc, which produces extremely high temperatures
- Disturbances in the heart's electrical conduction, which can cause changes in the heart's rhythm and even cardiac arrest
- Barotrauma from the acoustic and vibratory forces surrounding an arc blast
- Inhaled vapors, such as vaporized copper, released by an arc explosion[2]

The victim of the electrical accident is subject to the following types of injury:

- Burns
- Low-voltage contact wounds
- High-voltage contact wounds of entry and exit of the electrical current
- Injury to bone through falls, heat necrosis (death of tissue), and muscle contraction. (Shoulder joint injuries and fracture of bones in the neck are common injuries caused by muscle contraction.)
- Infectious complications
- Respiratory difficulties. (The tongue may swell and obstruct the airway, or vaporized metal or heated air may have been inhaled.)
- Injury to the heart such as ventricular fibrillation or cardiac asystole (the complete absence of electrical heart activity resulting in loss of blood circulation and lack of blood pressure and pulse)
- Injury to the eyes. (Cataracts from electrical injury have been reported up to three years after an accident.)
- Internal injury
- Neurological injury

Familiarity with the type of injuries possible in an electrical accident may help personnel to recognize their occurrence in an accident and treat the patient more effectively.

In most electrical accidents, the inability to diagnose the extent of injury at the time of admission to the hospital can delay the patient's treatment. Recovery can be enhanced by more detailed information about the accident, including amount of current, length of contact with current, and possibility of arc flash. Dr. Mary Capelli-Schellpfeffer of CapSchell, Inc., is a physician who specializes in treating electrical injuries and who has had a very positive impact on electrical safety research. She says, "The point is that when the doctors, nurses, paramedics and others better understand the injury scenario and the possible exposures a patient may have had, they can observe the patient with a heightened clinical suspicion for possible consequences."[3]

Electrical Accident Response Procedures

Sites should establish a training policy and plan to cover electrical rescue methods, approved rescue devices, and CPR training.

Immediate Actions

In response to an electrical accident, the following procedures should be followed immediately:

- **Remove the immediate hazard by turning off the power.** If you witness the accident, you must exercise extreme caution that you are not injured in addition to the accident victim. Always assume that the source of the electrical current is still energized unless you or another qualified person determines that the power has been turned off. Unless you are using insulated equipment (e.g., voltage-rated gloves, hot sticks, or a rubber blanket) to dislodge a victim, you must delay the rescue effort until the current can be interrupted.
- **Understand that speed is essential.** The victim's potential for injury increases with contact time. The resistance of the body is mostly in the skin. If the skin breaks down electrically, only the low internal body resistance remains to impede current flow.
- **Call for help.** Delegate someone else to get help, if possible. Be sure that an ambulance or emergency medical service is on the way.
- **Begin CPR.** If the victim's pulse or breathing has stopped, CPR is essential to avoid brain damage, which usually begins in four to six minutes. If CPR is needed, be sure assistance is on the way, but do not wait for help to arrive.

- **Be sure you and the victim are in a *safe zone*.** A safe zone is an area where a person cannot be in contact with any electrical source and where he or she is out of reach of any downed or broken wires. If the victim is unconscious, begin the CPR sequence.
- **Apply first aid to the victim.**
  - —If the person's clothing is on fire, remind him or her to drop and roll, or tackle him or her, if necessary, to smother the flames.
  - —Cool the burn with water or saline for a few minutes or until the skin returns to normal temperature. (For flash-burn victims, safety showers may be the best method, due to the possibility of widespread surface burns on the body.) *Do not attempt to remove clothing that is stuck to a burn.*
  - —Remove constricting items, such as shoes, belts, jewelry, and tight collars, from the victim.
  - —Elevate burned limbs to reduce swelling.
  - —Handle the victim with care, being aware that he or she might have broken bones or spinal injuries.
  - —Treat for shock: Maintain body temperature, do not give anything by mouth; administer high concentrations of oxygen, if available.
  - —Keep the victim warm and as comfortable as possible while awaiting transport to the medical facility. Cover him or her with clean, dry sheets or blankets. Cover burn wounds with sterile dressings or clean sheets.

Secondary Response—Helping the Caregiver

After the victim's immediate needs are met, the accident response personnel should consciously note as many details of the accident as possible. The details can help an accident victim's caregivers provide appropriate medical attention. It is especially important that hospital personnel know the cause of the victim's injuries. They need to know if the victim had contact with electricity or if arc flash caused the injuries.

The victim of electrical contact may suffer some surface burns where the current entered the body. He or she often suffers additional, less visible (internal) damage because of the path of the current through the body. The flash-burn victim is more likely to have greater evident burn damage on the surface of the body, due to the extremely high temperatures resulting from arc flash. He or she is likely to suffer first-, second-, and third-degree burns, especially on the face, wrists, ears, back of the head, neck, and ankles—any skin surface that is not adequately covered by protective clothing or equipment. In addition to burns to the skin, the flash-burn victim may have inhaled metal vapor—such as copper—or suffered adverse effects (such as damage to the eardrum) due to the pressure wave caused by arc blast.

The extent of injury to the victim of an electrical accident often is not immediately apparent. Injuries other than the obvious burns may occur, and caregivers must be aware of additional possible biological effects of electric shock. For example, electric shock may cause muscles to go into spasm with such force that bones are broken, and the path of the current flow through the body may cause other, life-threatening, internal injuries.

The extent of trauma to the body may be masked by the more readily apparent thermal effects of the injury. In most electrical accidents, the inability to diagnose the extent of injury at the time of admission impedes the patient's care. Treatment and recovery can be enhanced by more detailed information about the accident, including amount of current, length of contact with current, and possibility of arc flash. Recovery can be maximized by transporting the victim as quickly as possible to a burn center that specializes in electrical trauma.[4]

Information about the accident is imperative in the immediate care of the victim and in communication with the family. Facts collected immediately following the accident can be invaluable to the medical profession in aiding patient care.[5] In addition, complete information can help prevent similar accidents.

A major misconception is that accidents and injuries happen only to inexperienced personnel—not to those with many years' experience. Dr. Capelli-Schellpfeffer states, "If accidents don't happen to experienced workers, then why do I constantly hear experienced workers say, 'I never thought it would happen to me, but it did'?"[6]

Dr. Capelli-Schellpfeffer stresses that it is important to interview witnesses to accidents, if at all possible, to gain insight into how a victim can best be helped. She says, "Survivors of electrical trauma are often confused or disoriented after the accident. Therefore, why do we want to know *from them* what happened?" The more information caregivers can find out about the accident, the better care they can offer. Dr. Capelli-Schellpfeffer says that knowing the amount of voltage involved in the accident helps the physician to predict a patient's severity of injury and the length of the hospital stay. That information can be very helpful to caregivers as they talk to the patient's family.[5]

As part of the advance preparation for emergency response, site personnel should prepare a checklist, such as the one in Figure 14–1 on page 202, that provides detailed information about an accident.[6] The checklist was prepared by Dr. Mary Capelli-Schellpfeffer and the Electrical Trauma Research Team at the University of Chicago led by Dr. Raphael C. Lee. A completed checklist should accompany the victim to the hospital. Copies of the checklist should be readily available on site, and its existence should be communicated to all employees. Review of the checklist might be included as part of emergency procedure training for all personnel.

A Closer Look

Sam Barstow had worked for the Transco Corporation for 11 years.[7] The company had a good safety record, and Sam was a conscientious worker. The day of the accident started out much like any other. Sam was assigned to help a crew erect several steel girders for a new process building under construction.

It was almost lunchtime, but one large steel girder had to be maneuvered into place before the break. Sam was helping another employee, Walt, steady the steel girder. The load line on the crane was not quite vertical. The girder began to slip and contacted a 15,000-volt overhead electrical line. Sam felt the electricity enter his hands as he and his coworker were surrounded by a ball of light. Walt's clothing caught fire, and he was engulfed in flames. Sam experienced a brief period of paralysis and disorientation and injuries to his hands.

Sam and Walt were transferred by ambulance to the local hospital. Walt would live, but he was severely injured. Sam was hospitalized for a few days for treatment of hand and leg wounds, but he was able to return to work in less than a month.

Four years later, Sam was required by his company to undergo a psychiatric examination. Over the years, pain and weakness in Sam's hands had increased. He had been transferred by the company to a less physical work assignment because of his complaints, but Sam constantly felt stressed by the

This list should be a part of a site's emergency response plan for electrical injuries. A completed copy should accompany the victim to the hospital or treatment center if at all possible. The information will ensure the best possible evaluation and treatment by initial medical caregivers.

Name of injured person:_____

    1. When and where did the accident occur? _____

_____

    2. What was the victim doing at the time of the accident?_____

Yes No

_____

☐ ☐   3. Did the victim come in direct contact with electricity?

☐ ☐   4. Was an arc the source of electrical current exposure? If "yes," please explain.

_____

    5. What was the duration of exposure to electricity? _____

    6. Please identify the following as related to the incident:

       Voltage _____

       Capacity _____

       Source of electrical hazard _____

☐ ☐   7. Did the victim fall? If "yes," please explain. _____

_____

☐ ☐   8. Was the victim wearing protective or insulated clothing, safety boots, or gloves? If "yes," please explain. _____

_____

☐ ☐   9. Were others involved in the accident? If "yes," please explain. _____

_____

☐ ☐ 10. Before the accident, had the hazard been identified?

☐ ☐ 11. Did the victim seem dazed, confused, or lose consciousness at any point following the accident? If "yes," please elaborate. _____

_____

☐ ☐ 12. Did the victim require CPR?

☐ ☐ 13. Was the victim treated as if bones might be broken, especially in the neck?

☐ ☐ 14. Did the accident involve an explosion?

☐ ☐ 15. Did the accident occur in a closed space? If "yes," please elaborate. _____

_____

☐ ☐ 16. Did other hazards exist at the time of the accident, such as combustibles, heavy loads, moving or fixed machines, vehicles and equipment, or extreme ambient temperatures? If "yes," please explain. _____

_____

☐ ☐ 17. Name and telephone number of person who can provide further information about the accident events: _____

_____

Figure 14–1. Checklist for Victim of Electrical Accident.

Source: Adapted from questions prepared by Dr. Mary Capelli-Schellpfeffer and the Electrical Trauma Research Team at the University of Chicago led by Dr. Raphael C. Lee.

effects of the accident. He felt that he and Walt had been treated unfairly by the company. His marriage was in trouble, and he was afraid of losing his job and being unable to support his family.

Sam's case was reviewed by a multidisciplinary team. It was found that he was suffering from posttraumatic stress disorder, which included an inability to maintain attention and lapses in concentration, among other symptoms. Medical evaluation also found that he had decreased sensation in both hands, damage to nerves in his arms, and cataracts developing in both eyes.

Sam's case demonstrates that there is much yet to learn about the physical basis of neurologic and neuropsychiatric abnormalities in electrical injury. However, a psychiatric evaluation should be a part of an overall evaluation of electrically injured patients. Although psychiatric disturbance conplicates rehabilitation efforts, it is treatable. Long-term medical follow-up is also important, as physical problems resulting from electrical accidents can occur many months later. Information learned by initial caretakers about the electrical trauma is extremely important, not only in helping to diagnose immediate injuries, but in presenting a picture of care needed in the long-term recovery of the patient.

 ## Test Your Thinking

1. According to OSHA, personnel should check the contents of the site's first-aid kits as follows:
   a. Occasionally
   b. Once a month
   c. At least once a year
   d. Every other year
2. Barotrauma results from the following electrical accident results:
   a. Vibratory forces
   b. Acoustic forces
   c. Pressure from arc blast
   d. All of the above
3. What is the very first thing to do when an electrical accident occurs?
   a. Turn off the source of the electrical power
   b. Call 911
   c. Check to see if the victim is breathing
   d. Begin CPR

4. If an electrical accident victim's pulse or breathing has stopped, brain damage can occur within the following time range:
   a. One to two minutes
   b. Four to six minutes
   c. Thirty minutes
   d. If breathing has stopped, it is too late to save the victim.

Notes

1. OSHA 29 *CFR* 1910.269(b)(3).
2. Mary Capelli-Schellpfeffer, M. Toner, R. C. Lee, R. D. Astumian, "Advances in the Evaluation and Treatment of Electrical and Thermal Injury Emergencies."
3. Mary Capelli-Schellpfeffer, personal correspondence.
4. Jane G. Jones, "Report on Electrical Shock and Burn Treatment Captivates Audience at PCIC Safety Session."
5. Mary Capelli-Schellpfeffer, R. C. Lee, M. Toner, K. R. Diller, "Correlation between Electrical Accident Parameters and Sustained Injury."
6. Mary Capelli-Schellpfeffer, Christine Kaline, Michael K. Toney, John H. Mitchell, and Raphael C. Lee, "Partnerships for Electrical Safety."
7. The checklist in Figure 14–1 has been adapted from one originally developed by the Electrical Trauma Research Program, University of Chicago, Section of Plastic and Reconstructive Surgery. Used with permission.
8. This account is based on an actual incident. The names, including the name of the facility, have all been changed to protect those involved. Any similarity to actual names or facilities is strictly coincidental.

Answers: 1. (c), 2. (d), 3. (a), 4. (b)

■ Chapter 15

Developing and Implementing Procedures

What Is a Procedure?

Webster's Collegiate Dictionary defines the word *procedure* as a particular course of action or way of doing something. Some consensus standards use *practice* and *procedure* almost interchangeably. In this book, the authors distinguish between the two words.

Procedures are lists of sequential steps that identify actions thought to be necessary in order to accomplish a task. In most instances, procedures are written. A procedure can be verbal, but it is less likely to be effective. A practice is a normal, or routine, action or series of actions that a person always uses. A clear understanding of the distinction between these terms is important to understand consensus standard requirements. Unless the term is defined in a document, these understandings apply.

In some instances, the term *standard* is used interchangeably with *procedure*. A document that is called a procedure in one organization might be called a standard in another. However, legal implications sometimes crop up in how an organization uses these terms. For example, the term *standard* might be legally construed as mandatory, whereas the company's meaning was *guidelined* information. In most instances, indirect implications to the terms are simply unintended.

NFPA 70E, *Standard for Electrical Safety Requirements for Employee Workplaces*, 2000 edition, requires an electrical safety program that includes at least some procedures. OSHA standards require procedures to be in place. In fact, the term *procedure* is used over 700 times in the standards for general industry. OSHA standards frequently include generic sample procedures. NFPA 70E, Appendix C, suggests specific content for a procedure.

Most consensus standards assign the responsibility of defining and providing procedures to the employer. Employees generally are assigned the responsibility of implementing or following the procedure provided by the employer. However, an effective safety program cannot function that way. The program, including all procedures, must be a collaborative effort between the employer and the employee. Although the employer is legally responsible for providing procedures, the *people* involved, including people in both management and labor positions, have a moral responsibility to provide effective procedures. In most instances, a reasonably good understanding of terms exists within an organization. Communication problems exist when the discussion is among people with different backgrounds.

A procedure and a plan are essentially identical. However, a procedure generally covers a task that is routinely and frequently executed, whereas a plan generally covers a task that is one of a kind or executed only infrequently. A procedure might include a requirement for a detailed plan of execution. Equipment and circumstances vary among manufacturers, thus amplifying the need for a defined course of action.

Why Is a Written Procedure Important?

A written procedure is the best place to store important company information on how to safely execute a task. Electricians and technicians are curious and ingenious people; they continually seek a better (usually less labor-intensive) way to execute a task. If a written procedure exists, it can be easily modified to accommodate the new information. On the other hand, attempting to integrate any identified improvement into the work process is much more difficult. Written communication is much more effective than verbal communication only. A written procedure should be included in all site training efforts. Either a single procedure or a portion of a procedure can serve as the point of discussion for regular crew meetings.

A written procedure serves as sound guidance for an employee as he or she executes a task that might be somewhat unfamiliar. Not executing a task for a period of months can mean that a worker has less-than-complete recall of how the task should be executed. Detailed steps in a procedure can help to refresh a worker's memory. If the employee (or trusted coworker) participated in generating the procedure, he or she will be more likely to embrace any safety requirements specified in the procedure. As a tool, a procedure is invaluable to employees.

A recently reviewed/revised/reissued procedure provides a sound legal argument for both an employer and an employee. The existence of a current procedure illustrates the intent and effort of an employer. Even the existence of an out-of-date procedure illustrates intent and effort. Likewise, the lack of a procedure illustrates the lack of employer interest. In litigation, attorneys may use the state of procedures to illustrate employer intent. That point can be either positive or negative for an injured person. A current, up-to-date procedure can be used to defend an OSHA citation. Sound reasoning that is recorded in a current procedure can be effective in mitigating such citations.

When to Write a Procedure

In some organizations, a procedure develops from plans being written several times for the same task. It is unimportant whether a unique plan is written for a task or a procedure is used for the same task, provided that the plan or procedure is safe and effective. Some organizations might write a unique plan to remove a circuit breaker for maintenance every time the task is performed, thus writing a plan several times a year. Logic suggests that writing a plan more than three times for the same task should be sufficient to establish a routine way to complete a task. Maintaining a procedure requires much less effort than generating the procedure in the first place.

If a task involves complex actions or requires significant exposure to safety hazards, then a written procedure is usually necessary. In many instances, a procedure covers controls and permit requirements as determined by the degree of hazard exposure. A procedure is the best place to define and implement these requirements.

As discussed earlier in this chapter, both OSHA and consensus standards include requirements for a procedure to exist and be maintained. In these cases, a written procedure must exist.

How to Write a Procedure

Different organizations take different approaches to writing a safety procedure. In some organizations, the safety managers (or their designated representatives) jealously guard the task of writing and stewarding safety procedures. After all, they believe that they were trained and hired for that purpose. In other organizations, the electrical superintendent (or engineer) writes the electrical safety procedures. In still other organizations, the electrical superintendent (or engineer) writes procedures as directed by the safety manager. Although none of these processes is necessarily wrong, none produces the best procedure.

It is important that the people who must implement the procedures have significant input in the process of developing the content. The people who do the work are the best source of information on how to execute a particular task. They are the ones who are most exposed to hazards.

Safety professionals are most expert in safety processes. The safety manager is generally the person who deals with the aftermath of an injury. The manager is usually most informed about legal issues associated with preventing injury.

Technical experts (engineers and supervisors) are usually most informed about consensus standards, and they may even participate in the consensus process. They also tend to converse with colleagues in other companies and in other environments. Usually, it is a technical expert who is involved in specifying equipment and directing the design and installation of electrical equipment. Technical experts tend to develop and direct maintenance practices and frequency.

All of these employees—workers, safety professionals, and technical experts—have significant information that must be combined in order to generate the best safety procedure. Experience has shown that when all three types of people work together, the product is more easily implemented and produces the best results.

The authors recommend that a single person produce the first draft of an electrical safety procedure. After the draft is generated, a second person with similar experience should review and critique the draft. Shortly after the first review, the reviewer and the drafter should discuss the first draft and make changes until both are satisfied with the resulting second draft.

The second draft should then be reviewed and critiqued by the technical expert and a member of the safety organization. Shortly after the second review, a committee comprising the drafter, the subject matter expert, and the safety professional should thoroughly discuss the second draft. Agreed-upon changes should be made, resulting in a third draft.

The third draft should be circulated to all interested people for review and comment. Any suggested change that results from this review should again be discussed in the committee and appropriate changes made to the procedure. The final procedure then must be published for use.

What to Put in a Procedure

The content of a written procedure varies, depending on the objective of the procedure. For example, a safety procedure includes information on safely executing a task, whereas a maintenance procedure provides information on maintenance activities and tasks that are necessary to effectively maintain equipment. Both types of procedures have similar components; however, this chapter discusses safety procedures rather than maintenance procedures.

Elements

Each safety procedure should address the following specific elements:

- Scope, or purpose, of the task
- Extent, boundary, or limits of the task
- Employee qualifications and number of employees involved
- Type of hazards expected
- Limits of approach
- Safe work practices to be used or avoided
- Personal protective equipment required
- Materials and tools required, permitted, or prohibited
- Any special precautionary techniques
- Electrical drawings to be referenced
- Equipment details that might surprise the worker
- Sketches or photos of unique features

Although one or more of the listed elements might not be necessary for any specific procedure, all of the elements should be considered and a conscious choice made regarding the need to address each issue. Each of the elements is discussed in the text that follows.

Scope or Purpose of the Task

The procedure should first identify the task or work activity that is covered by the procedure. The scope or purpose statement should provide sufficient information for the reader or user to know specifically what the intended task is.

Extent, Boundary, or Limits of the Task

The reader or user must be able to determine the intended boundary of the work or task addressed by the procedure. In other words, the reader or user must be able to determine what is *not* covered by the procedure. Frequently, knowing what is *not* covered is at least as important as knowing what *is* covered.

Employee Qualifications and Number of Employees Involved

If specific employee qualifications and training are not required, it is unlikely that a safety procedure is required. Therefore, employee qualification is important. The procedure should describe what worker qualifications are required, as well as how many employees are necessary to safely execute the task.

Type of Hazards Expected

The procedure should identify what hazards are associated with the task. Performing a hazard analysis and then cataloging the results of that analysis in the procedure are very important. In fact, products of the hazard analysis should be more than the electrical hazards associated with the task. For example, danger of falling, contact with high-temperature pipelines and equipment, or co-occupancy problems in the area should also be identified. If unusual or unique methods of exposure exist, the procedure should also identify them.

All hazards should be considered. An injury from a fall or high-pressure steam can be devastating to a worker and his or her family.

Limits of Approach

The procedure should address the limits of approach if exposure to shock or arc flash is a possibility. Knowing the limits of approach is critical. If the procedure does not identify the limits of approach, it must instruct the user where to find that information. In most instances, a procedure will neither contain limit-of-approach information nor refer the user to the location of the information. However, a one-time use plan probably would contain the limits of approach. (See Chapter 2, "Electrical Hazards.")

The procedure should consider whether physical barricades are warranted while the work is in progress. Sometimes a physical barricade is needed and, at other times, only a sign is needed (with no barricade). If the procedure cannot firmly require or reject a physical barrier, then it should contain guidance so that the user can make an individual decision.

Safe Work Practices to Be Used or Avoided

If specific safe work practices must be used to minimize hazard exposure, the procedure should contain that information. For example, experience might have shown that installing temporary ground sets reduces the exposure to shock. If this is the case, the procedure should require the installation of grounding sets. The important issue is that the procedure user has the necessary corporate guidance contained in the procedure. If some work practices must be avoided, the procedure must also prohibit the practice. For example, experience might have shown a propensity for workers to forget to remove temporary ground sets when work is complete. The procedure should address these conditions.

Personal Protective Equipment Required

The procedure should recommend personal protective equipment (PPE) that will mitigate exposure to any hazard that remains while the work is being executed. If PPE is covered in a different procedure, a reference to that procedure might be all that is needed.

Materials and Tools Required, Permitted, or Prohibited

If special or tools and/or materials must be used for the task at hand, the procedure must identify them. The procedure must also identify any materials or tools that are prohibited. If specific material characteristics are desirable, the procedure should identify those characteristics.

Any Special Precautionary Techniques

If executing the task requires unique exposure to a hazard or exposure to a unique hazard, the procedure must identify any associated requirement. For example, if the task requires interaction with a laser, then special eye protection might be necessary.

Electrical Drawings to Be Referenced

The procedure should include a list of reference drawings that might be needed to execute the work. In almost all cases, a current single-line diagram is needed. Wiring diagrams or schematic diagrams might also be needed. This section of the procedure should tell the user which drawings are needed and where to find them.

Equipment Details That Might Surprise the Worker

It is possible that the equipment on which the task is to be performed has been modified since it was installed. Perhaps the worker has not yet been exposed to this particular model of equipment, or perhaps this particular equipment is experimental in nature. The procedure should provide any needed warning.

Sketches or Photos of Unique Features

If sketches or photographs of the internal arrangement of the equipment exist, the procedure should include each photo or sketch. If the reference information is located somewhere else, the procedure should identify that location.

Where to Keep the Procedure

For any written procedure to be effective, it must be available to everyone who may need to use it. All procedures should be available in the field. They should be easily accessible to all. If a manual is produced, it should hold a prominent spot on the desk of the first-line supervisor, and a copy should be kept in the shop.

Some companies have all active procedures available in computer systems where users are expected to log in and either read or print the procedure as needed. In some instances, a user is comfortable with accessing the information on a computer screen. In other cases, the user feels a need to have a copy of the procedure in his or her hand. The "environment" that exists within an organization has a bearing on how a person feels about using a computer to access a procedure. The real issue is the ready availability of the procedure when needed.

Some organizations have expended significant amounts of money to generate very strong procedures. Usually these organizations design and produce attractive binders and store the procedures in those binders. Site managers and safety departments place copies of the safety manual in visible locations in their bookcases, where they stay.

Other organizations write their procedures and bind them together with staples. They make many copies of the "safety manuals" and distribute them to all employees. The safety manuals become dog-eared from use. Usually, these safety manuals are much more effective than the above scenario where attractive manuals stay on the desk.

Procedures are of no real value unless they are used, discussed, changed, and updated as user experience is gained. The procedures are only effective tools when they are in the hands of the users.

A Closer Look

Joe Schwartz was an electrician for Brown Enterprises.[1] He had been an electrician for 29 years, most of the time with Brown. He was very vocal about how work should be performed. His long experience established him as the informal leader of the younger crew.

In an attempt to be more competitive, Brown Enterprises had recently begun to shift to multiple-craft supervision. Joe's current supervisor, Al, was experienced in the mechanical crafts and was familiar with the mechanical aspects of the electrical craft. Al was unaware of how much the rest of the crew respected and embraced Joe's ideas.

A new OSHA rule had just been promulgated, and the site safety manager was aware that wholesale revision of the electrical procedures was in order. The new rule received a great deal of attention in the management ranks. The main point of discussion with the new rule was that a hazard called arc flash was working its way into the regulatory process.

The safety manager asked Al to review the current electrical safety procedures, paying particular attention to what needed to be done to protect people from this "new" hazard. In order to fulfill the request, Al sought help from Brown Enterprises' engineering staff. Because the engineering staff had been keeping abreast of IEEE's developing knowledge on arc flash, an engineer agreed to write a procedure for Al.

Al was confident that the engineer was familiar with how to protect people. He accepted the draft procedure and relayed it to the safety manager for action, without discussing the procedure with his electricians. The safety manager reviewed the procedure and placed it into the system for issue. No attempt was made to gain the "buy-in" of the electrical workers. The gist of the procedure was that switchman's jackets and switchman's hoods were to be worn at all times when a door was opened on any electrical equipment.

When the procedure "hit the street," Joe was very vocal in his denunciation of the procedure. In all his 29 years as an electrician, he had never seen so much overkill. He had seen electrical explosions, but nothing that warranted such an outfit. He simply was not going to wear the prescribed gear. Anyway, he complained he wouldn't even be able to take an ammeter reading while wearing the hood. He doubted that he would be able to see the voltmeter. He simply refused to follow the new procedure.

On the day that the protective clothing was delivered to the site, Joe was called to troubleshoot a balky motor. He grabbed his tool belt and thought briefly about the new procedure. Again, he made a conscious choice to ignore the procedure.

When Joe arrived at the work site, he noted that the overload button in front of the motor control center unit was missing. He put on his leather gloves and reached for his screwdriver to release the door fasteners. After the door fasteners were released, he moved the switch handle to the "off" position. As soon as the switch handle began to move, a broken spring on the handle flew off and made contact between phases A and B. The resulting arc flash and blast blew the door open. A fireball and molten metal flew out of the now open door. Joe's glove prevented significant burns on his hand, but the fabric in the sleeve of his shirt melted onto his arm. Joe went to medical and was transported to a nearby hospital, where the burn was treated.

Although the PPE identified in the procedure would have prevented injury, other PPE would have been sufficient for Joe's exposure, as well. Had Joe participated in the production of the procedure, he probably would have been following the procedural requirements and avoided an injury.

 ## Test Your Thinking

True   False

☐   ☐   1. Effective procedures must be available in a hard-copy form in the plant manager's office.
☐   ☐   2. A procedure should identify all hazards associated with the subject task.
☐   ☐   3. Only the safety department writes effective procedures.
☐   ☐   4. A procedure and a plan are similar.

Note

1. This account is based on an actual incident. The names, including the name of the facility, have all been changed to protect those involved. Any similarity to actual names or facilities is strictly coincidental.

Answers: 1. (false), 2. (true), 3. (false), 4. (true)

■ Chapter 16

The Changing State of Electrical Safety

After electricity came into use, for many years electrical accidents and injuries were accepted as inevitable. Even today, too many electrical and managerial professionals accept significant exposure to hazards as a cost of doing business. However, *all* accidents truly can be prevented.

Across the United States and in other industrialized nations, many individuals and organizations are devoted to improving the statistics of electrical incidents, accidents, and injuries. They strive to improve accident statistics in a very personal way—by preventing them, if possible, but also by learning from incidents and improving the lot of the victims. As more becomes known about the nature of electricity, more also becomes known about how to prevent accidents. New methods and philosophies are being developed that can help management to protect workers and, most importantly, help workers to protect themselves. In addition, physicians and medical personnel are working to improve victims' chances for survival and lessen the effects of electrical shock, flash, and burns.

The Current State of Industry

In the current industrial "downsizing" environment, many companies are eliminating middle and senior managers. At a time when employers are being asked to do more with fewer resources, attention to electrical safety in the workplace becomes more important than ever. According to Dr. Mary Capelli-Schellpfeffer, however, "By developing the obligation of coworkers to themselves and each other, management can move beyond the expectation of the OSHA rules that suggest that occupational safety and health is an employer duty."[1] Industrial employees must come to realize that even though their employer puts safety procedures in place and complies with regulations, the employees' ultimate safety depends on their own awareness, personal principles, and refusal to take risks and engage in unsafe practices. Employees can and must take responsibility for their own safety.

The recognition by industry of arc flash as a hazard is a promising new development that has changed the state of electrical safety. Because arc-flash incidents are now recognized as other than simply "electrical burns" (i.e., thermal burns), more is being learned about both prevention and treatment of arc-flash injuries (see Chapter 2, "Electrical Hazards").

Awareness and Education

A key element in preventing electrical accidents is education. The more that people understand how and why accidents occur, the more aware they will be of the possibilities both for incidents and for ways to prevent their occurrence. "Spreading the word" about electrical safety involves educating and training employees. Creating an awareness of hazards and how to keep those hazards from causing accidents is a goal worth pursuing.

As understanding of hazards and related injuries increases, employees become less willing to take risks, supervisors become less willing to assign tasks that involve risk, and managers become more willing to expend money to reduce exposure to risk. As employers train and educate employees, employees themselves begin to accept personal responsibility to avoid exposure to hazards. Education makes possible the fact that speaking up to question work patterns is acceptable to the community. Everyone can and should "be smart" about electrical safety.

Petroleum and Chemical Industry Committee of the IEEE/IAS

The Petroleum and Chemical Industry Committee (PCIC) of the Institute of Electrical and Electronic Engineers (IEEE) and the Industry Application Society (IAS) works to disseminate information on electrical safety. This group has held an annual conference for nearly fifty years. An entire general technical session of each annual conference, led by the PCIC Safety Subcommittee, is devoted to electrical safety. Many excellent papers presented at the conference have reported cutting-edge advances and research in electrical safety. The Safety Sub-committee of the PCIC also holds an annual safety workshop that provides, according to its image brochure, "a forum for people to meet and exchange ideas for preventing electrical accidents and injuries in the workplace . . . to accelerate the dispersion of understanding and knowledge of ongoing improvements in the practical application of codes, standards, technology, and implementation methods that reduce the risk of electrical injuries."

National Electrical Safety Foundation

The National Electrical Safety Foundation (NESF) is an organization in the United States that is devoted to promoting awareness of the importance of electrical safety in the home, school, and workplace through education. The foundation was established in July 1994 through the initial collaborative efforts of the Consumer Product Safety Commission, the National Electrical Manufacturers Association, and Underwriters Laboratories Inc. Using the theme "Plug into Electrical Safety," the NESF aims to improve electrical safety in the home. Materials include "Home Electrical Safety Quiz," "Outdoor Safety," and "Promoting Electrical Safety in the Home, School, and Workplace." Another important effort of the NESF is promotion of May as National Electrical Safety Month. Materials are available from the National Electrical Safety Foundation, 1300 N. 17th Street, Suite 1847, Rosslyn, VA 22209; telephone (703) 841–3211.

May Electrical Safety Month

Many industrial sites have joined in promoting May as Electrical Safety Month. A monthlong emphasis on electrical safety awareness is one way to heighten awareness of workers, especially those not normally involved in the electrical discipline. The annual campaign offers electrical associates a unique opportunity to bring greater awareness to home, work, and public electrical safety issues. Sites choose special campaign themes and promote them through special programs, speakers, posters, contests, newsletters, skits, and other original ideas. Such a promotion has proven to be both effective and popular among employees. Members of the electrical safety team and those most knowledgeable about electrical safety usually lead the efforts.

Technological Advances

In a democratic society, profit and costs generally drive changes in technology. Other reasons also cause change, but the reason for change is almost always economic in nature. Companies are either economically successful, or they are not. Those that are unsuccessful go out of business, regardless of their products' usefulness.

In reality, products that impact electrical safety are no different from any other product. Only customer demand drives the development of new technology. Corporations spend money to develop products that are either in demand or are anticipated to be in demand. Without purchase orders, a company cannot market a safety product that reduces employee exposure to electrical hazards.

Arc-resistant switchgear is an example. Several companies know how to produce a product that will consistently reduce employee exposure to hazards associated with an arcing fault. Installation standards permit this equipment to be installed but contain no requirement for the equipment. Arc-resistant switchgear is more expensive than other equipment. Some of this equipment is being purchased and installed; however, the quantity of these installations is very small. Organizations might have high value for protecting people from arc flash and still not understand the degree of their employee exposure.

Research and Testing

Research is expensive. Testing is expensive. It stands to reason, then, that a manufacturer should expect to recoup the investment costs associated with the research. However, without adequate tests, no realistic way exists to predict how an item will function when conditions are not normal.

In most cases, tests involve standards that are written essentially by the manufacturing community. Manufacturers are in the best position to understand their own line of products. However, it is also true that product tests on which manufacturers agree do not address all of the user issues. Therefore, can the tests be adequate without input from users?

In many places, OSHA requires products to be tested and labeled by a nationally recognized testing laboratory (NRTL). An NRTL is a test facility that is recognized by OSHA representatives in accordance with the protocol in 29 *CFR* 1910.7, Appendix A. This appendix details the process for applying and becoming a recognized NRTL.

Some testing is privately funded. In some instances, several corporations will pool resources to design, execute, and analyze the results. Generally, findings of these tests are published in a public forum. The public forum then makes the information available for further use by individuals. These efforts normally are scrupulously monitored in order to escape any appearance of business collusion.

It is important to remember that almost all testing of electrical equipment is associated with equipment integrity or installation integrity. Very few tests are intended to provide any indication of how people will interact with the equipment. Electrical equipment that is used to deliver energy should be considered potentially dangerous to people. Although tests are normally conducted to determine the ability of the equipment to survive in a bolted-fault condition, essentially no tests are conducted to determine if a person can adequately interact with the equipment in any condition. No tests consider if an average person can understand the labels or indicators. No tests are conducted to determine how the people/equipment interface will react in an arcing-fault condition.

Much of the basic research conducted in recent years has been under the auspices of the Electric Power Research Institute (EPRI), a consortium of electrical utilities. The primary mission of EPRI is to conduct research in issues that are important to electrical utilities that comprise its member companies. Dues and assessment of member companies provide funding for EPRI. The member companies then share the results of the research efforts.

Growth of Standards

As the business climate expands to include binational, trinational, and other trade arrangements among international bodies, pressures increase for broad, generally accepted standards. Standards-developing organizations feel the pressure to "merge" requirements or adapt requirements to be increasingly "friendly" to other national requirements.

Standards that are produced in an international working environment are increasingly important. Standards that are accepted across international borders are also increasingly important. Because standards influence trade, either positively or negatively, pressures on standards-developing organizations are growing.

In many instances, national standards from different countries take different approaches to address the same engineering problem. In the United States, for example, the English system of measurement has been used throughout its history. In Europe, on the other hand, the metric system is used. Consequently, wire sizes are manufactured and sized in different units. A system to prevent explosions in flammable and combustible liquids and gases generally relies on sound, but different, engineering methods. Wiring methods vary. Installation requirements vary. Committees that write standards are becoming increasingly aware of and familiar with other engineering approaches to solving the same problem.

Electrical hazards remain the same, regardless of the location or type of government or the ability of the electrical workers. The method and degree of exposure to a hazard might vary from one locale to another. However, electricity *always* follows the laws of physics.

Personal protective equipment developed to align with experience and perceived need does vary between locations. Like other electrical products, PPE was developed over the years, as necessary, to meet the demand of the purchasing community. As standards recognize a "new" (previously unrecognized) hazard, the purchasing community tends to accept the new requirement and purchase the equipment.

Medical Advances

As discussed earlier, electrical power distribution and generation have had an extreme impact on modern industrial society. The trauma caused to victims of electrical accidents and its clinical management has also influenced modern medicine as caregivers strive to treat patients. Says Dr. Raphael C. Lee, "Physicians now encounter complex injuries resulting from electrical forces, which range today from commercial frequency electric power to ionizing irradiation. With the development of new power electronic devices operating at new frequencies, physicians are compelled to broaden their understanding of electrical trauma."[2]

University of Chicago Burn Trauma Center

One institution that has added considerably to the understanding of electrical trauma is the University of Chicago Burn Trauma Center. The Electrical Trauma Research Program is directed by Dr. Raphael C. Lee. Dr. Lee is uniquely suited to lead such an effort. He is trained as a general surgeon and a plastic surgeon, and he also received an SC.D. in engineering from the Massachusetts Institute of Technology. His team's research in the area of electrical trauma has added immeasurably to the knowledge base and understanding of electrical injury.

One such area on the cutting edge of research is the work done by the team on cell membrane damage. The team demonstrated that much of the damage from electrical burns resulted from electroporation, the opening and enlargement of pores in cell membranes due to high-intensity electric fields rather than the heating of tissue by the passage of current. Electroporation allows essential ions to flow freely through the membranes, eventually resulting in cell death.[3] Dr. Lee and his colleagues showed that the use of surfactant compounds injected into the cell membranes within a 30-minute period could successfully seal and repair the membrane, thus reducing the effect of the injury on the victim.[4] Additional research has been performed by Dr. Mehmet Toner of the Harvard Medical School, Massachusetts General Hospital, and the Shriners Burn Institute.

Several organizations have supported the work of the University of Chicago team and other research in the treatment of electrical injuries, including the Amoco Foundation and the Electric Power Research Institute, Empire State Electric Energy Research Company, New York State Electric and Gas Company, Niagara Mohawk Power Corporation, Northeast Utilities Services Company, Public Service Company of Oklahoma, Public Service Electric and Gas (NJ), and Wisconsin Electric Power Company. The support of these organizations has helped to benefit everyone. However, research continually requires economic support, especially the support of industry.

Incident Reporting

Assessing the number and severity of electrical incidents in the field is very important in helping to understand and prevent injury. Counting electrical incidents is very difficult, however. Employers are inconsistent in their effort to collect information and report incidents. Peer pressure in workplace cultures actually tends to discourage incident reporting. Management is not rewarded for reporting, nor is drawing attention to unsafe acts appreciated. Employees are expected to conduct their work in a safe manner, and results to the contrary may cause unwelcome consequences.

The FAR Project

The purpose of the FAR (Fax Anonymous Registry) Project for Electrical Incidents is to collect information about the scope of electrical events in diverse environments. The collection process is designed, in a nonthreatening way, to address the disincentives in reporting electrical incidents. The project, started and supervised by Dr. Mary Capelli-Schellpfeffer, ended its first official reporting year on September 30, 1999. The documentation of electrical incidents was adopted from the DuPont Scheme for Electrical Incident Recordkeeping, which was shared at the IEEE PCIC Electrical Safety Workshop at San Antonio, Texas, in 1997. Incidents are recorded by witnesses or others involved and reported anonymously so that information can be collected. Names, places, and companies are considered unimportant in the collection of incident data.

Project participation is invited. Incident reporting is always anonymous. The reporting form may be found on the CapSchell, Inc., Web page, www.CapSchell.com. The fax number for the project is USA 262–552–2902. Readers can call confidentially and speak with Dr. Mary Capelli-Schellpfeffer at USA–773–960–5802.

Incident/Injury Database

As previously stated, an incident/injury database of electrical incidents and accidents is important in understanding and preventing accidents. The National Institute for Occupational Safety and Health (NIOSH) has agreed to use the information collected by the FAR Project in its database. Currently, NIOSH publishes a report entitled "Worker Deaths by Electrocution: A Summary of Surveillance Findings and Investigative Case Reports." The publication reports electrocutions as a major cause of death based on data from the NIOSH National Traumatic Occupational Fatalities surveillance system and includes NIOSH recommendations focusing on prevention. The free publication is available from Publications Dissemination, EID, National Institute for Occupational Safety and Health, 4676 Columbia Parkway, Cincinnati, OH 45226.

 # A Closer Look

On an October Thursday, Ryan went to work anticipating the sixth World Series game.[5] Houston was down three games to two. Ryan had four tickets to the game and would be leaving Beaumont at 2:00 P.M. to drive to Houston with his family. He had planned to work in the morning before leaving for the game. He was as excited as his sons.

Ryan was a senior electrician at a large chemical plant near Beaumont. He had "paid his dues" at the plant and felt that he had earned an easy morning. He was assigned to the electronics crew in the maintenance organization.

The solid-state drive room was air conditioned in an effort to remove heat generated within the power electronics in the drives. Each section of the drive line-up had an exhaust duct at the top of the unit and a louvered door with an air filter on the inside of the door. This solid-state drive was old equipment. The equipment was physically very large. Electrically, the equipment was very large, as well. The transformer supplying the rectifier was rated at 2,000 kVA. The rectifier was close-coupled to the transformer, so there was no secondary overcurrent protection on the 480-volt bus.

The supervisor knew that Ryan was going to the game and wanted him to enjoy the outing. Ryan was spared all the heavy work on this Thursday. Instead, his work task was to change all of the air filters in the drive room. Ryan knew that the filters were all the same size: 12 by 18 inches. The filter material was man-made fiber, constrained in shape by aluminum screen, and the filters were light in weight. Ryan knew that the drive equipment would be running, but he also knew that would cause no problem. The filters were always changed while the equipment was running. All he had to do was open the door, change the filter, then close the door. There were no interlocks to defeat.

Ryan picked up the filters from the storeroom and put the box on his tool cart. Yessir, he would have an easy morning. He proceeded to the drive room and opened the first door. He changed the filter and discarded the dirty filter in the trash container nearby.

Ryan moved on to the second unit. He sure hoped that Houston would win the game. He hoped someone in his family would catch a foul ball. He knew the bat boy and would be able to get the ball autographed. As he reached for the door handle to open the door, he heard a noise. It seemed to be coming from the large rectifier at the other end of the room.

Ryan moved closer to the rectifier. Yes, the loud humming was coming from inside the rectifier. Instinctively, Ryan opened the door of the unit to look inside. He had done that a thousand times before. This time, however, there was a problem. He never learned the source of the humming. The explosion was immediate, as soon as the door was moved. A short circuit occurred somewhere on the bus from the transformer. Ryan's cotton clothing caught fire. No one else was in the room, but other workers came quickly when the noise from the explosion reverberated around the facility.

The pressure forces from the explosion blew Ryan against the adjacent wall. His lifeless body was transported to the hospital.

After an in-depth analysis of the incident, the chemical plant relocated the filters to the outside of the door. In the future, changing filters would become a job for two people. Arc-flash-resistant clothing would be provided for all employees who were required to enter the room. No rectifier or transformer door would be opened with the equipment running.

That action was too late to help Ryan. He never knew that Houston won the ball game.

 ## Test Your Thinking

1. Which of the following methods is/are most important in preventing electrical accidents?
    a. Employers must comply with regulations by providing protective equipment for their employees.
    b. Employees must be responsible for their own safety.
    c. Only answer (a)
    d. Both (a) and (b)
2. When hazards and exposure correlation with injuries are better understood, which of the following things happen?
    a. People become less willing to take risks.
    b. Supervisors become less willing to assign tasks that involve risk to people.

     c. Managers become more willing to expend money to reduce exposure to risk.

     d. All of the above

3. Which of the following statements is/are true?

     a. Electrical hazards remain the same, regardless of the location or type of government or the ability of the electrical workers.

     b. Electricity *always* follows the laws of physics.

     c. Methods and degrees of exposure to a hazard never vary from one locale to another.

     d. Answers (a) and (b)

     e. Answers (a), (b), and (c)

4. Which of the following contribute to low reporting of electrical incidents?

     a. Peer pressure in workplace cultures

     b. Drawing attention to unsafe acts is not seen as being in employees' best interests.

     c. Employees are expected to conduct their work in a safe manner, and results to the contrary may cause unwelcome consequences.

     d. All of the above

Notes

1. Mary Capelli-Schellpfeffer, "What Can Management Do?"
2. Raphael C. Lee, *Injury by Electrical Forces: Pathophysiology, Manifestations, and Therapy.*
3. "Improving the Diagnosis and Treatment of Electrical Burns."
4. Raphael C. Lee, Adam Myerov, and Christopher P. Maloney. "Promising Therapy for Cell Membrane Damage."
5. This account is based on an actual incident. The names, including the name of the facility, have all been changed to protect those involved. Any similarity to actual names or facilities is strictly coincidental.

■ Chapter 17

Working with Another Discipline

Operating a facility and producing a product require work in several disciplines. Even in very small companies, more than one discipline is required. For example, a small contractor requires some effort in sales and marketing. The same company has both administrative and managerial work in addition to craft work. The organization's degree of success depends on successfully executing work in these different disciplines.

A small company might not have sufficient resources to employ personnel trained in each needed discipline. The only real option is for a person who is trained in one discipline to assume responsibility for another.

As larger companies attempt to become more competitive by reducing the number of employees, economic and business pressures force the organizations to hire multidisciplinary employees and supervisors. Although most people have the ability to learn and execute work in another discipline or craft, those who already have the necessary training in and understanding of a second or third craft are very unusual.

In an expanding organization, members of the organization feel good about a supervisory assignment that covers more than one discipline. In most cases, the multidisciplinary assignment means that the organization is expanding and becoming more successful.

However, in a large organization, a supervisory assignment for employees of more than one discipline might cause employees to feel threatened. Regardless of how the multidisciplinary assignment is announced, employees tend to conclude that their discipline or craft is less valued after the initiative is started. As a result of feeling threatened, the employees may be more withdrawn and less cooperative, which amplifies the multidiscipline supervisor's lesser degree of training.

When the multidisciplinary assignment involves the electrical discipline, exposure to electrical hazards increases. One result of this increase in exposure is a commensurate increase in injuries. How people feel about their work in an organization has a significant bearing on how they react to safety initiatives.

The Safety Professional

Some companies employ a person who is trained to implement, operate, and maintain a safety program. That person is dedicated to creating a safe work environment. He or she is a safety professional. In fact, some organizations are large enough to need several safety professionals. In many instances, the safety professional is called the *manager of safety*, or *safety manager*. The safety professional might be trained in another engineering discipline but fill the safety manager role for the organization.

One characteristic of human nature is a desire to feel powerful. An organization's safety manager generally reports to top management, suggesting that the position is a powerful one. This organizational structure suggests to all members of the line organization that the safety manager can influence the operation of the site or organization. In some instances, the safety manager is reluctant to share any power that is inherent in the way he or she reports up the line of the organization, especially if that reporting goes straight to the top. Frequently, however, any reluctance to share the inherent power is a false impression based on how the safety manager acts and reacts.

Safety professionals usually are expert at developing and implementing an overall safety program. In fact, they normally have a good understanding of hazards, how people are exposed to them, and how that exposure might be avoided or minimized. Safety professionals expend a great deal of their efforts following requirements that are developed in an environment external to the company.

Electrical safety hazards are vastly different from hazards associated with other kinds of energy. Previously unrecognized hazards are making their way into standards. Few safety professionals come from the electrical discipline. Because of the complexity of the growing knowledge of electrical hazards, safety professionals need support from electrical professionals.

Electrical professionals are trained to understand electrical energy. In most cases, electrical energy distribution and utilization border on "black magic" to untrained members of the community. Electrical professionals are reluctant to take time to increase understanding of their coworkers in another discipline. By not taking the time to build a coworker's understanding, however, an electrical professional exhibits the same need to feel powerful that other professionals demonstrate.

Electrical professionals understand electrical hazards and how people are, or might be, exposed to them. Their reluctance to share that understanding leads to safety programs that are less effective than they could be.

Sharing responsibility for the safety program could be interpreted as sharing power that is normally inherent in the training and organizational assignment. Safety managers might view themselves as less powerful. Advice provided by electrical professionals would be open to criticism by someone from another discipline. The safety professional might "lose face" if he or she were unable to defend his or her position before the criticism. If the objective of both the safety professional and the electrical professional is to impress the boss, neither professional is likely to share their organizational power.

On the other hand, if the personal objective is to prevent injury from an electrical hazard, both the safety professional and the electrical professional have a much better chance to achieve the objective by working together. The safety professional seeks and accepts advice from his or her electrical colleagues. The electrical professional seeks and accepts advice from his or her colleagues in the safety office. When people freely admit that they don't know it all, they strengthen the lines of communication and open the door for improvement and education of all concerned.

As suggested in Chapter 10, the electrical safety program should be one part of an overall site safety program. The safety manager should rely on the electrical community to develop and maintain electrical safety procedures and practices. At the same time, the electrical community should rely on the safety professional to direct the overall safety program, including the electrical safety portion of the program. It is important that all efforts to control or manage exposure to electrical hazards be coordinated with all safety efforts. An electrocution is no more devastating to a family than a fatality from a fall. The only real difference in injury from these potential results is the method of exposure and the mechanics of prevention.

A Supervisor from Another Craft

Almost every person believes that his or her craft or chosen discipline requires greater skill and knowledge than every other skill or discipline. This trait is common among electricians, electrical engineers, lawyers, plumbers, doctors, reporters, politicians, and astronauts, as well as all others. Believing that his or her chosen career requires great skill and knowledge fulfills an inner desire to feel important.

Everyone wants to feel personally important. When a person is promoted or appointed to supervise other people, the supervisor develops expectations of himself or herself. Other people also develop expectations of the supervisor. These expectations almost always include an anticipation that this person has more knowledge than the people who report to him or her.

When the supervisor has expertise in the craft or discipline of those people reporting to him or her, the expectations tend to remain. Although the supervisor's expectations of himself or herself are moderated, some degree of the expectations endures. However, when the supervisor is from another craft or discipline, employees tend to be suspicious of the supervisor's technical qualifications in their craft. To them, the supervisor cannot possibly comprehend the intricacies of the craft or discipline being supervised. This point seems to be especially true when the craft or discipline is electrical and the supervisor is perceived to have no experience in the electrical discipline.

As the industrial environment becomes more competitive, there is greater incentive for a corporation or company to move to multicraft supervisors. Significant economic pressure causes the company to maximize the contribution of each employee. When a full crew of people is not required in a particular craft or discipline, either the supervisor is responsible for people from another craft or the supervisor's workload is shortened. On a large site, workers might be supervised by a person from another physical area (minimal supervision) or by a person from another craft.

A single person might be supervising employees from two or three different disciplines. The obvious difficulty is that a supervisor might assign workers to tasks where exposure to hazards is elevated. The worker probably executes the task without advising the supervisor about the elevated hazard exposure. The supervisor and the line organization don't know about the elevated degree of exposure until something happens.

The employer should provide electrical hazard training for all supervisors. All members of the line organization should become familiar with all known electrical hazards. Hazard awareness training should also be provided for all employees who are or may be exposed to electrical hazards in their work assignments.

For the electrical safety program to be effective in a work environment influenced by multidisciplinary supervisors, people who are or may be exposed to electrical hazards must be provided with more training than would otherwise be necessary. The impact of effective procedures is much greater in this type of work environment. Frequent and effective hazard awareness training is critical.

Workers must feel empowered to talk freely about their work. Each worker must be active in his or her own safety. Supervisors must be very open to comments and criticism. *Any* worker comment must be accepted as authentic. Criticism of supervisors and managers is a norm. How the supervisor or manager reacts to the criticism has a great impact on the willingness of the worker to offer comments in the future. Sometimes such comments might make the difference in preventing a tragic accident.

A Closer Look

The Baylor Corporation had a film process facility in San Antonio, Texas.[1] The plant had been very profitable for many years. However, with the influx of new competitors in the film business, the Baylor plant was feeling pressure from the new competition. Baylor's facility was old and less efficient than the new facilities of the competition.

The plant manager and her staff were under pressure to reduce costs— to become more competitive. At the suggestion of human resources, the plant manager elected to implement the corporate early retirement program at the San Antonio plant. Although some of the personnel reduction would obviously come from the worker ranks, the majority of potential savings would come from reduction in the ranks of managers and supervisors. The plant would reorganize around the core of supervisors and managers that remained after the lucrative retirement opportunity.

Juan Silva, an electrician, had been at the Baylor plant for 27 years. In fact, he was the first electrician hired, going to work even before the plant went into production. Juan had helped to check out and start up the facility. He knew the plant like the back of his hand.

Juan was a good listener. He had great respect for his immediate supervisor. In fact, his immediate supervisor had been one of his best teachers. Much of Juan's technical electrical knowledge had been gleaned from the three supervisors he had worked for at the plant in those 27 years.

Like Juan, Bryan had worked at the Baylor plant for over 25 years. Bryan began his career as a machinist. However, for the past 18 years, Bryan's assignment was operating production equipment. As the first-line supervisor, Bryan's contribution was deemed significant as the reorganization unfolded. Plant management determined that his talent was underutilized. Juan's small crew was reassigned to report to Bryan.

The elimination of the electrical supervisor made Juan feel that the plant staff had less value for the electrical craft—his craft. Bryan had been a supervisor for a long time, and he was used to being in charge. He felt that if he was supervising electricians, he was going to tell them what to do, not the other way around. After all, maintaining electrical equipment was little different from maintaining mechanical equipment. Listening to Bryan as he lined up people for their daily tasks and watching how Bryan acted, Juan was well aware of how Bryan felt. Juan felt devalued.

One Tuesday morning, Juan was assigned to investigate a problem with electrical panel 1-AC, located on the wall in the lunchroom. This panel supplied air-conditioning equipment that was common in the area. The air-conditioning was needed only for comfort, but it was an important creature comfort for the employees. All of the air conditioners were window units that were installed in shops and change areas.

Juan was trying to determine if the problem was in the transformer—perhaps with a connection within the enclosure. He had the front cover off the transformer. He had not locked out the transformer because he was only looking at the terminals. Bryan came rushing to the transformer and asked Juan to go over to the control room and see what was wrong with the production line. Juan told Bryan that the transformer cover was open, but Bryan did not really understand. Bryan did not recognize any particular hazard. He didn't even ask if the transformer was locked out. Juan told Bryan that he would go as soon as the cover was back on the transformer, but Bryan said that the production line was down. The more Juan tried to argue with him, the more insistent Bryan became. Juan was to go immediately. Juan picked up his tools and went to the control room.

Half an hour later, the contracted janitorial service entered the lunchroom, where the transformer was located. The contract employee began mopping the floor with a wet mop. The contractor coordinator approached the employee cleaning the floor. The employee stopped to listen, and the wet end of the mop when straight into the open transformer. The employee screamed and jumped out of the way. He had received a severe shock. He was sent to the hospital, where he was interviewed and released. The contractor employee was very fortunate that the wooden handle of the mop in his hand was relatively dry.

The Baylor Corporation was lucky, too. The incident was a near fatality. Shortly after the incident, an electrical supervisor was appointed. It was Juan Silva.

 ## Test Your Thinking

True   False

☐   ☐   1. The electrical safety program should be one part of an overall site safety program.
☐   ☐   2. Safety professionals normally need support from subject-matter experts.
☐   ☐   3. Production supervisors make good supervisors for electrical workers.
☐   ☐   4. Electrical workers do not need supervision.

Note

1. This account is based on an actual incident. The names, including the name of the facility, have all been changed to protect those involved. Any similarity to actual names or facilities is strictly coincidental.

Answers: 1. (true), 2. (true), 3. (false), 4. (false)

■ Chapter 18

Planning and Communication

Planning the job is one of the most important safe work practices. A safe and effective plan plays an extremely important role in the ability to manage exposure to an electrical hazard. An effective plan can overcome other shortcomings in the electrical safety process. If a device is known to be problematic, a plan should be devised that can limit any resulting exposure to an electrical hazard. If the qualifications of an individual are less than desired, the plan can account for assigning that person to less critical work tasks. Although mitigating other potential problem areas is potentially possible, that result is unlikely.

What Is a Plan?

Sometimes a person arrives physically at the job site while he or she is mentally in another place. When a family member or another loved one is in trouble, it is probable that the worker continues to think about the loved one, either consciously or unconsciously. In fact, many different reasons might cause a person's mind to stray from the task at hand. Looking forward to an exciting event or reviewing an argument or disagreement might interrupt a person's ability to focus on a task. A plan helps to focus a worker's attention more closely.

A plan is an agreed upon series of steps or actions that are necessary to accomplish an objective. It describes how the task will be accomplished. A plan can be any of the following:

- A written document
- An unwritten work practice
- A consensus document
- A consensus requirement

A plan can be simple or complex and prepared in any of the following ways:

- It might be written specifically for a task at hand.
- The plan might be a simple discussion with a coworker or a colleague.
- The plan might also be stopping the work process for a few minutes and mentally going through the work steps before starting the task.

In the best of circumstances, as well-planned work unfolds, unexpected complications might arise. In fact, the whole idea of troubleshooting is to find and correct an unknown problem. If two or more tasks are attempted while the troubleshooting task is under way, people involved in the multiple tasks may have trouble coordinating with one another (having the same plan). As an unexpected condition is recognized or identified, the work process should be stopped and replanned. Many accidents and injuries occur when unexpected conditions occur and the workers continue to follow the original plan. The work should stop, and a new plan should be generated.

An electrical task is sometimes influenced by work going on in another craft or discipline. For example, routine maintenance on a motor control center or distribution panel that is located in a process area may be affected by a task involving thermal insulation maintenance in the same area. It is not unusual for two different disciplines or crafts to independently plan work for the same area and time, especially when both tasks are associated with the same production process. Each discipline should have independent plans; however, the individual plans must be coordinated. Communication among the disciplines is the key. *The most important reason for having a plan in place is that everyone involved has the same idea of what is going to happen and knows how to accomplish his or her part of the task in relation to everyone else's part.*

The Job Line-Up

A supervisor might have a specific idea in mind and fail to communicate what he or she is thinking during the job line-up. More frequently, the supervisor is thinking about the mechanical steps necessary to execute the task, with little thought about personal safety. Even when the supervisor has specific safe work practice steps in mind, he or she might forget to mention them during the job line-up.

In the utility industry, holding a tailgate discussion prior to beginning a task is a common practice. The idea of the tailgate discussion is that a plan of action is discussed so that each person involved in the work holds the *same plan*. Even in those instances where an overall written plan is produced, the tailgate discussion still must ensure that each worker is working toward the same plan. The concept of a tailgate discussion is that the discussion takes place just before the job is begun—not on the first of the month or two days before—so that the plan is fresh in everyone's mind as they start the job.

In other segments of industry, similar discussions also occur. In general industry, producing an overall plan of how the work is to be accomplished is a relatively common practice. However, the plan may not be shared with each worker. Of course, it is the worker who is exposed, or potentially exposed, to an electrical hazard. Even though supervisors frequently discuss hazards associated with the work, workers may derive a different work plan. A tailgate discussion should be held prior to beginning the work. The tailgate discussion is the last opportunity to make sure that each employee and supervisor has the same plan in mind.

Sometimes a job involves many different tasks over the course of many days. Several different crews may be involved in the work. An overall plan must be developed so that all involved can understand how multiple tasks and the work of multiple crews fit together to accomplish the entire job. Standing procedures are likely to be involved in executing the work. The overall work plan should identify all procedures that will be involved in executing the work. It is important to review the procedures that will be involved to ensure that they are adequate for the specific need at hand. All referenced procedures and the entire work plan should be made available to employees. Then, at the beginning of each discrete task, a tailgate discussion should be held to discuss any hazards associated with the immediate work task.

A project schedule cannot be completed until some basic planning has been accomplished. Initial planning is likely to be done by a scheduler or an engineer. Unless the person doing the initial planning has performed similar work, the plan might be somewhat sketchy. Experience has shown that involving a worker in the initial job planning provides information that otherwise would not be considered. The plan should include the experience of each craft involved in the work process. If technicians are to be involved in the job, they should be involved in producing the plan. If electricians are involved in the work, they should also be involved in the planning process.

Procedures should be supplemented, as necessary, to implement intent. Consensus standard requirements for lockout and tagout establish a need for each person to be in *control* of all energy sources. Unless each machine or process in a facility has its own discrete lockout/tagout procedure, a generic procedure probably should be generated and supplemented with additional information for each lockout. Additional information should be provided concerning the location of appropriate energy-isolating devices. No lockout should be executed without a supplemental list that shows all involved energy sources and energy-isolating devices. (Such a list is also very helpful to monitor who installed locks on which energy-isolating device.)

Communicating the Plan

Once the plan is produced, the next important step in the process is to communicate the content of the plan to everyone who needs to know. And, every person involved in the work process needs to know. Each one should be familiar with his or her role as defined within the plan. Workers also need to know what their role is *not*. It may be possible to communicate a minor role for an employee within the context of a tailgate discussion. However, it is likely that each crew member involved in the work process reviews the overall plan in a common meeting. The meeting can be held in a conference room, in a shop, or in the work area. The discussion should be sufficiently detailed and communicated well enough for workers to understand their complete role in the overall plan. An opportunity must also be provided for two-way discussion and questions to be asked and answered.

A single person-in-charge should be appointed, and everyone should know who that person is. The person-in-charge should be assigned the responsibility of monitoring the completion of each step in the plan.

The plan probably will not be executed exactly as initially planned. When something goes wrong or an unexpected situation is uncovered, the person-in-charge should stop the progress of that step in the process. He or she should consider all implications of the new information. If implications for another step in the plan are involved, that work should also be stopped until each step affected by the new information is replanned. Usually, this type of new information changes the degree of exposure to a related safety hazard. Any such change must be analyzed, and any appropriate mitigating steps must be put into place.

A plan that has been well considered with all necessary thoughts can still be ineffective. When an electrical task requires more than one person, everyone participating in the work process must be following *the same plan*. The inability to communicate effectively can result in two or more different understandings by different people. To avoid this possibility, discussion and feedback are necessary. Each step of the plan must be discussed in sufficient detail to establish a common understanding among all the workers involved in the task.

A plan is not a plan when different people understand the steps in different ways. Incidents and injuries are frequent results of different people having different plans in mind for the same task. This is especially true when two people are working together on the same task.

 # A Closer Look

For 13 years, Randy Williams and Xena Foster had worked together at an automobile assembly plant.[1] Their working relationship was strong from the beginning. Because Randy was physically large, the coworkers became known as Xena and Hercules. After working together for so long, each of them could almost read what the other was thinking.

The automobile assembly plant made wide use of plug-in bus to distribute power to the various machines scattered throughout the facility. In many cases, 5,000-amp ventilated bus connected a transformer to its secondary switchgear that was located several feet away.

On one Friday morning in March, the supervisor told Xena and Hercules about a problem with the robot welder on line 2 in the body assembly area.

The supervisor said the operator reported that the robot just "quit working." The machine would do nothing.

Xena had a hunch: The 100-amp switch feeding the unit had been acting suspiciously for some time. She had trouble operating the switch handle the last time the machine had a problem. The supervisor had no plan for troubleshooting the switch. He indicated that he did not really care how they "fixed the problem." He just wanted the production manager off his back. Xena and Hercules understood.

Xena found no voltage at the incoming lugs in the robot control panel. Hercules went to get a ladder so they could reach the switch on the plug-in bus. He retrieved a fiberglass ladder from the control room and met Xena in the production area below the plug-in switch. The plug-in bus with the problem switch was mounted at an elevation 5 feet above a section of the ventilated bus.

Xena helped Hercules place the ladder into position and held it while he climbed to secure the top with the rope. Both Hercules and Xena wanted to prevent a fall from the ladder. Hercules was wearing his fall protection harness, so they were ready to check out the problem switch.

After a second cup of coffee, Xena put on her tool belt and climbed the ladder. She moved the operating handle to the "off" position, but the knife blades did not move. She had trouble opening the cover of the 100-amp switch as she stood on a rung of the ladder. Hercules asked her to come down and let him try.

Hercules climbed up the ladder. He decided to improve his leverage by stepping off the ladder onto the ventilated bus, which was at a slightly lower level. When he stepped on the top of the ventilated bus, his weight caused the cover to bend slightly. That bending of the cover was enough to cause it to contact the energized bus within the enclosure. The ventilation holes in the bus directed the force of the blast and the thermal energy directly upward, right onto Hercules.

Hercules had not taken time to tie the lanyard on his safety harness, and he fell onto the floor, right in front of Xena.

Xena called for help from the medical department and went back to try to offer comfort to Hercules. Hercules was transported to the hospital. He had received burns over 70 percent of his body. He remained alive for the next three days. However, Hercules would never recover from the burns and fall. Neither would Xena.

Test Your Thinking

1. A plan may be or include which of the following?
   a. A written document or an unwritten work practice
   b. A consensus document or requirement
   c. A discussion with a coworker
   d. All of the above
2. Which of the following is the most important reason for having a plan in place before beginning work?
   a. To keep electricians from doing the jobs that laborers should do
   b. So that everyone involved has the same idea of what is going to happen and knows how to accomplish his or her part of the task in relation to everyone else's part
   c. So that breaks can be scheduled appropriately
   d. So that workers can plan their vacation days accordingly
3. Which of the following benefits can be secured by holding a tailgate discussion before beginning the task?
   a. Making sure that each employee and supervisor has the same plan in mind
   b. Allowing the supervisor to relate to the workers any specific safe work practice steps he or she has in mind
   c. A discussion of hazards associated with the immediate work task
   d. None of the above
   e. (a), (b), and (c)
4. Lockout/tagout is an important part of isolating electrical equipment to be worked on. Which of the following should be secured before lockout is executed?
   a. A copy of the OSHA regulations for each person involved
   b. A supplemental list that shows all involved energy sources and energy isolating devices
   c. Knowledge that each person involved in the task is in control of all energy sources
   d. Both (b) and (c)
   e. All of the above

Note

1. This account is based on an actual incident. The names, including the name of the facility, have all been changed to protect those involved. Any similarity to actual names or facilities is strictly coincidental.

Chapter 19

Effective Auditing of the Workplace

One of the most effective tools in the safety manager's tool kit is the audit. Analysis of the audit results can reveal the effectiveness of the overall safety program to a manager or supervisor. Depending on the overall work environment, negative connotations may be associated with the term *audit*. Employees may fear repercussions when deficiencies are uncovered, and some might think that an audit would just increase their workload. Others just object to being "watched." However, involving employees in the audit process may overcome negative feelings associated with the word. For the audit results to yield the most benefit, all members of the organization, including both workers and supervisors, should embrace the auditing process. The audit might be called a survey or an inspection if the negative connotation to the word audit is severe.

Defining the Objective

The objective of a safety audit is to produce data that can be analyzed to gain insight on how to improve the safety aspect of the work environment. Audits can be conducted of any aspect of a work site or job. To establish a benchmark, a routine safety audit should be established. On large sites or projects, the site or project might need to be subdivided, with two or more teams conducting an audit simultaneously. The same team, however, should perform the complete routine audit so that the measurement parameters are equal. If more than one team is involved in performing the audit, different measures or observations are likely to be used to define a "violation."

Preparing for the Audit

The site should have a general understanding of what the audit team will be recording. For example, a violation might be an action or condition that is not aligned with a procedure. An audit might reveal poor or unsafe work practices, such as the following, which should be considered violations:

- A blocked fire extinguisher
- Standing on a handrail
- An open cover on an energized electrical switch

General agreement should be reached by the audit team that each condition or practice found causes increased exposure to a hazard before it is cited as a violation.

Just as hazards differ by degrees, different degrees of violations exist. A violation that might result in a fatality is more significant than a tripping hazard. Therefore, more significant violations should receive more attention.

Some organizations assign numeric values to the different degrees of violations, then use those assigned values to calculate a site or area violation frequency rate (VFR). The VFR can be normalized to a per-employee average basis and used for comparison. Contests among areas at the site facility can compare the VFR of each area. Contests between or among sites and contractors can help to increase interest in safe work practices, good morale, and a positive safety environment.

If a contest has been established, the "winner" should receive an award. Awards have some short-term benefit and should be considered. The economic value of any award is less important than its existence.

Conducting the Audit

The audit team should include people with different expertise and experience levels. Each "violation" must be recorded, along with the disposition of the violation. If the violation was corrected while the audit was being conducted, that fact should be recorded. If the violation poses no immediate hazard exposure, the violation can be corrected later. It is important that all violations be corrected and that employees are aware of these corrections. Violations should never be used to embarrass individuals or teams of employees.

Each audit or survey should have a central focus. The focus should change from one audit period to another. However, all observed violations should be recorded, whether in the focus area or not. The audit should be routine and regular. Employees should expect their work area to be audited at the appointed time.

The audit team should assemble and begin as scheduled. If more than one team is required, each team should begin to walk through the assigned areas.

As the audit team progresses through the plant or area, each member of the team should observe employees as they work in addition to conditions in the work area. The team members and other employees should talk freely with one another; however, observed violations should be recorded. The recorded violation must be independent of employee names, but some attribute must be recorded so that the violation can be identified. The idea is to avoid embarrassing a person or persons but still be able to correct the violation.

Following Up after the Audit Is Completed

The team's recording secretary (the person keeping notes) should transfer any recorded information to a form for distribution. The VFR should be calculated, and that number should be publicized to the organization. Any violation that was not corrected while the audit was going on should be corrected expeditiously. A time limit should be assigned by which all observed violations are to be corrected. Each location should be revisited after the time limit has expired, and the correction should be verified.

If the site or employer has a VFR contest, audit teams or site management should congratulate the winning area and award chosen prizes. The identity of the winning area should be broadly publicized.

Comparing the VFR from one week or audit period to another provides a measure, or benchmark, to determine whether or not site safety is improving. Assigning each observed violation to a category provides a measure to determine where funds or emphasis should be placed for the greatest impact.

One important result of the audit should be that work areas are better organized. Workers tend to function more efficiently in an area that is free of such obstacles as tripping hazards and clutter. They seem to be better able to focus their thoughts on their work when the work area is reasonably well organized. One result of the audit is that at least once each audit period (at least weekly) the area is cleaned and organized.

Conducting Interim Audits

Each time a supervisor walks through an area, he or she should review the work conditions and employee work practices. Each time a manager walks through an area, he or she should review *if* or *how* people are exposed to hazards. Any exposure or potential exposure to a safety hazard should be corrected through the line organization. Using the line organization to correct potential exposure to a hazard emphasizes that personal safety is important.

Auditing Training and Documentation

Consensus standards require industrial sites to maintain several types of records. OSHA requires several others. These records should be audited at least once a year. Training records frequently are maintained in the personnel file of each employee (an excellent idea). Each job description should include training requirements necessary for a person to execute the job effectively. The job description should include a listing of safety training requirements, including the dates when the training was successfully completed.

In some instances, employee records are kept as data in a computer program. Executing an audit of these records may become academic, as the computer can schedule employees for training. If the organization then implements the training as scheduled by the computer program, no deficiencies will be found by the audit. However, external agencies (e.g., OSHA) may have a legal right to review audit information for training records. The authors recommend that a hard copy of the audit information be maintained on site. The audit record should be no more than 12 months old.

If an audit reveals any deficiency, each deficiency must be dealt with effectively. A record of how the organization corrected each deficiency must be noted on the audit record.

Auditing Behavior

The engineering community deals with hard information and facts: exact engineering data and pure logic. It is not unlike an average person. Most managers, engineers, and others deal with hard, visible facts. When an incident or accident occurs, the normal first reaction is to add another guard, increase the ability of the equipment to survive without an injury, add pressure relief vents to the equipment, and so on. This concept is viable and an important element in preventing injury.

When a modification of equipment doesn't seem to solve the problem, the next usual action (or reaction) is to add personal protective equipment (PPE) so that when the equipment does fail, an injury can be avoided. Examples of this action are adding shoulder harnesses for every person in the car or adding more layers of arc-flash protective equipment.

About one-third of all incidents and injuries could be prevented if the concepts discussed above were "clinically" implemented. Perhaps two-thirds of incidents and injuries would still occur, since approximately that same percentage is associated with behavior.

Behavior can sometimes be influenced by the physical state of the workplace. As an area becomes more congested, people tend to spend more time thinking about the congested workspace and less time thinking about the task being executed. If exposure to an electrical hazard is rare, an employee's behavior might be influenced by exposure to the hazard. It is possible to analyze the physical audit from a behavioral perspective, and site supervision should attempt to perform such an analysis.

The most effective method of assessing behavior is to talk with the employees. This "audit" should not involve a committee of people. A single individual talking one on one with another person has the best chance of eliciting and understanding why employees act as they do.

Discussions with individuals should never be associated with a person by name. Instead, any suggested "reason" for an identified behavior should be recorded anonymously. A system, or process, should be put into place within the organization that addresses issues with negative impact.

Improving the Program and Planning for Potential Improvement

The result of effective auditing is one additional source of information that can be used to improve a safety program. However, the organization can benefit only if supervision reacts to the results of the audit. Correcting all violations identified by an audit has at least two benefits: Potential hazards are removed, and all members of the organization are reminded that the company is committed to the safety program.

 # A Closer Look

Charlie Jordan was a journeyman electrician for Caruthers Construction.[1] Caruthers had been awarded a contract from PSG Corporation to install a feeder from substation 3-J to a new distribution panel for a powerhouse environmental project. The distribution panel had been set into place, and cable tray had been installed. The cable tray was supported on an existing pipe bridge near a paved road inside the plant.

The feeder cable had been designed as three large, aluminum conductors with a PVC jacket over the composite assembly. Caruthers had a reel of copper cable in stock in the warehouse that was larger than the cable specified by the design. The plant sought permission from the engineer to use the larger cable instead of purchasing new cable. Because the proposed cable would more than adequately handle the current demands, the engineer gave permission for the cable substitution. Both PSG Corporation and Caruthers Construction saved a few bucks with the substitution, and "saving a few bucks" was always a priority of both companies.

Both Caruthers and PSG routinely monitored expenditures. Their prime intent was getting the job done. Neither company paid much attention to work processes or work areas. The construction supervisor pushed his crew to keep them working and, in the process, permitted the work area that had been isolated with barricade tape to become very cluttered. There was no place for the large cable reel in the immediate work area.

When the cable for the new feeder was delivered, the construction supervisor requested that PSG enlarge the work area. The area would have to increase the next day anyway, when the mobile crane arrived to help with the cable pull. Organizing the work area and removing trash would have created sufficient space for the cable, but that approach would have required the trash removal service to make another visit to the site, thus increasing costs.

The following day, the mobile crane was positioned adjacent to the barricaded area. The plan was for the crane to lift the reel above the existing substation so the cable would flow into the cable tray as it exited from the reel. The plan was put into motion. The cable was rigged and lifted by the crane. Because of the small, congested work area, the crane had to lift the reel and cable from the side and swing them toward the front of the crane. The outriggers were in place.

The reel and cable cleared the ground, and the crane operator began to swing the load toward the front of the vehicle. Just as the crane began to swing, the left front outrigger punched through the earth into an underground sewer pipe. The crane tipped over onto its left side, dropping the reel onto a roadway. The crane operator was able to stay on the crane until it came to rest. No one was injured. The cable and reel could have dropped onto an energized unit substation, but, luckily, they didn't. The load could have crushed a person in a car traveling along the roadway, but, luckily, it didn't. Only the crane, cable, and sewer pipe were significantly damaged.

Removing the trash and clutter would have enabled the crane to be placed as originally planned, but the extra visit by the trash removal service was expensive. The cable and reel originally specified weighed much less than the substituted large copper cable, but that fact was not realized. The original plan was not reviewed after the cable substitution, thus time was saved (and a few more dollars). Both PSG and Caruthers viewed the incident as an unusual accident. Neither recognized that the incident could easily have been avoided. The site had saved money, but the companies had narrowly escaped a real tragedy. In addition, now the sewer pipe had to be replaced. The cable was destroyed, and the crane would need to be repaired. Was saving money worth the risk?

Test Your Thinking

True   False

☐   ☐   1. Involving employees in the auditing process may cause negative feelings to be associated with the word *audit*.

☐   ☐   2. The objective of a safety audit is to produce data that can be analyzed to gain insight on how to improve the safety aspect of the work environment.

☐   ☐   3. An audit can result in a work area being cleaner and better organized.

☐   ☐   4. Audits should be kept secret while they are being conducted, and the results should not be made known to employees.

Note

1. This account is based on an actual incident. The names, including the name of the facility, have all been changed to protect those involved. Any similarity to actual names or facilities is strictly coincidental.

Answers: 1. (false), 2. (true), 3. (true), 4. (false)

Chapter 20

Personal Safety Principles

Webster's Collegiate Dictionary defines a principle as "a comprehensive and fundamental doctrine or a code of conduct." Almost every person has some fundamental values that he or she is not willing to overlook. Every productive member of society has someone or something that he or she holds in great esteem. These fundamental values vary widely from one person to another, just as behavior varies widely.

If a principle is a fundamental code of conduct, it is logical that if a person thinks about what he or she is doing, then that person will act in a way that aligns with his or her principles. Defining and understanding electrical safety principles attempts to draw each person's fundamental doctrine into his or her conscious thought processes.

The first need, then, is to identify those basic principles, to generate an understanding of how those principles are "connected" to something that the person holds in great esteem, then to define and implement a process that will continue to bring attention to each person's conscious thinking.

Identifying Electrical Safety Principles

Several effective processes may be used to identify and build initial understanding of each person's principles. One that has been effective in the past is to assemble a small group of people (at least 6, but no more than 10) who have high regard for electrical safety. The personal characteristics of the assembled group of people are important. They should be willing to share their thoughts in a noncritical brainstorming session. They should be willing to listen and try to understand any thought expressed by anyone else.

The willingness of the assembled group to openly discuss basic beliefs with each other is extremely important. Discussion should begin with open dialogue about that point. If unanimous agreement cannot be reached, the degree of success will be less than desired.

The session should continue with a discussion of what the group is trying to do—that is, to identify their personal safety principles. The intent should include identifying basic values of the average member of the company. Those present should try to understand who or what the average person holds dear, such as spouse, child, or parent. The idea is to begin to develop an understanding of how an average person's spouse or child is or may be affected by electrical safety hazards in the workplace.

After the initial understanding is achieved, the discussion should continue by trying to identify and begin to understand the assembled group's principles that keep them safe from electrical hazards. This discussion is likely to be a little ragged and unsettling as the group wrestles with the idea that each person's basic value system indeed has a bearing on whether or not he or she is willing to accept exposure to an electrical hazard. The group must continue discussions until the members have collected a set of words or statements that describe the principles that guide their daily behavior.

This discussion can be draining for group members who have participated freely in the discussion. At this point, group members should relax and take a break from the heavy discussion. However, they should not return to their normal daily routine. A communal lunch break is helpful.

After the break, the group should reassemble and talk about the words or statements that were agreed upon earlier. Now the discussion should focus on decreasing the number of words. Short, catchy phrases are much easier to remember than longer phrases. The length of the statements is not really important. What is important, however, is that the electrical safety principles that are the completed product are relatively easy to remember. The principles should also reflect the values that most members of the organization can rapidly understand. (See Figure 20–1.)

We Have Them—Now What?

Although the meeting can be dismissed, the members of the group cannot be. Those people should become disciples of the principles. Now it is important to get everyone in the organization to hear and begin to understand those principles.

- Plan every job
- Anticipate unexpected events
- Use the right tool for the job
- Use procedures as tools
- Isolate the equipment
- Identify the hazard
- Minimize the hazard
- Protect the person
- Assess people's abilities
- Audit these principles

Figure 20–1. Electrical Safety Principles.

Electrical safety awareness programs should be designed around the principles. In fact, more than one program is in order. Electricians might need one type of training, whereas other workers might need another. Managers certainly need a different training program. The whole company should be trained to understand the principles.

The electrical safety awareness training program should discuss incidents and injuries that someone within the organization has experienced. The program should include some humor as it looks at one or two incidents in a realistic way. The discussion of incidents or injuries should relate to each of the electrical safety principles and illustrate how applying at least one of the principles would have avoided the injury or incident. The incident discussion should illustrate how the injured person's spouse, child, or parent was affected by the injury. The overall intent of this portion of the training program is to evoke an emotional connection with the injured person.

Following the electrical safety awareness training, it is important for communications and discussions to relate to the principles. Job line-ups, safety audits, and incident reports and analyses should relate to the principles. Managers and supervisors should include the principles in discussions with employees at every opportunity. Generating pocket-sized cards that include the electrical safety principles and providing them to each person is an excellent option. If the cards are generated, each person should be encouraged to carry one with them and review them from time to time.

Electrical Safety Principles—An Example

The following set of principles was generated in the environment described above. The group in this example determined that the safety message produced should be placed in the public domain so that it might be used by all. Therefore, no permission is needed to use them. Since that time, many versions of the principles have emerged. One modified version is included in NFPA 70E, *Standard for Electrical Safety Requirements for Employee Workplaces,* 2000 edition.

Plan Every Job

Every significant job or task is made up of several discrete steps. Each step that is required to complete the job should be considered independently. Thinking about each step that is physically required to finish the assignment provides an opportunity to identify if it is possible to execute the work. Sometimes it is not. If it is possible to remember all the steps required to execute the job, it is not necessary to write down each step. If several steps are needed, each step should be written on paper. The idea is that the person performing the task has a clear picture of all physical steps that are required to execute the job.

If more than one person is involved in the work, all people involved in the work process should have the same plan in mind. Many incidents occur because more than one plan exists.

Anticipate Unexpected Events

Once a clear picture of all the steps in the job is in mind, the worker should consider what could go wrong as the plan is executed:

- What would happen if the wrench or screwdriver slipped?
- What would happen if someone bumped into the worker as he or she executed the task?
- What should the worker do if the arrangement behind the cover or door is different from what was expected?

If an unexpected event occurs or an unexpected condition is found, the work must be stopped and a new plan established that considers the "new" information.

Use the Right Tool for the Job

People are ingenious. In most instances, they will find a way to accomplish an assigned task. Frequently, the ingenuity requires the person to misuse a hand tool, such as the following instances:

- Pliers are sometimes used instead of a wrench.
- A wrench is used that is the wrong size.
- A screwdriver tip is used as a shim.
- A ladder is used when a platform is really needed.
- A multifunction meter is used to test for absence of voltage when a single-function meter is needed.

Only tools that are designed for the purpose should be used. Injuries occur when the wrong tool is used.

Use Procedures as Tools

A procedure is essentially the same as a written plan. In fact, a procedure is a job plan that is previously prepared. Some jobs or tasks are repeated often. A procedure is a written plan that eliminates the need to write a plan each time the job is performed. Because the procedure was prepared previously, it is important to recognize that the equipment might have changed or deteriorated. For example, the overcurrent protection might be incorrectly sized. Therefore, any previously prepared plan could be missing some very important information. The procedure should not be considered as law; it should be a tool that can help the worker to consider thoroughly all steps of the job.

Like any other tool, a procedure should be reviewed from time to time and revised as necessary to keep it sharp and functional. Also, like any other tool, if it is not available when needed, its value disappears.

Isolate the Equipment

Troubleshooting, maintaining, or modifying electrical equipment normally involves potential exposure to shock, arc flash, and blast. All sources of energy should be removed from the equipment. Personal locks should be installed. Personal tags should be installed. If all energy sources are removed from the equipment, no possibility of injury from those sources remains.

Any electrical equipment that is opened for work should be physically isolated. This point is especially important when an electrical source remains energized within the equipment. A barricade or other warning signal should be installed to ensure that unqualified people remain outside the limited approach boundary and the arc-flash boundary.

Identify the Hazard

In the vernacular of most consensus standards, no requirement is needed for employees to be able to identify a hazard. However, electrical hazards are very difficult to see. Each step of a plan should be reviewed to identify any personal hazard someone might be exposed to while executing the task. This step of the analysis must consider all hazards, not only electrical hazards.

Minimize the Hazard

If it is not possible to completely remove each safety hazard, all possible approaches to minimize exposure to any hazard must be taken as the task is being executed. For example, all doors must be kept closed, if possible. All possible energy sources must be isolated. The number of people who are or may be exposed must be limited. Insulating materials must be installed to minimize the duration of exposure to shock or minimize any potential arcing fault.

Protect the Person

If all hazards cannot be removed, any person within the boundary of exposure *must* be protected from any potential result of an electrical malfunction. If a person is within the arc-flash boundary, he or she should be wearing protection from any potential degree of release of an arcing fault. If the person is working where any physical contact with an energized component is possible, he or she should be wearing shock-protective equipment. If he or she is working from an elevated position, the person should be wearing fall protection, as appropriate.

Assess People's Abilities

One key element that can ensure that an incident or injury does not occur is examination of the qualifications of the worker who is planning, supervising, and executing the work. It is critical that each of these people have a good understanding of electrical hazards. The ability to avoid an incident is heavily related to the degree that the workers understand the hazard. As the degree of understanding increases, the ability to avoid an incident also increases.

Audit These Principles

It is important for each person to review the text that makes up his or her electrical safety principles. The principles should be changed, as necessary, to enable the person to continue to implement willingly the thoughts expressed by each safety principle. Change probably will not be necessary, but each person should consider the possibility.

 # A Closer Look

Craig Miller had worked at Penguin Manufacturing Company (PMC) for 23 years.[1] He considered himself to be a good employee. He also considered himself to be an honest, caring individual. Craig had worked on the power distribution crew for several years, and he was very familiar with the double-ended substations prevalent at PMC.

PMC was in the chemicals business. Craig had noted that over the past eight to ten years, PMC business had declined about 15 percent. The chemicals industry was becoming more competitive. Craig wanted to do whatever he could to help the PMC business environment.

The decreased business had caused PMC to reassess the job assignment structure in the Dallas facility. Some jobs were lost in the reassessment, but all personnel losses were voluntary. Since the reorganization, PMC seemed to be regaining some of its lost sales momentum.

Substation BA-5A1 contained two transformers and two primary feeders. The transformers were close coupled on opposite ends of the equipment line-up. The 2,400-volt load cables could be supplied by a circuit-breaker position in either the top or the bottom compartment. Doors in the rear of the equipment were hinged. Each door uncovered not only the load cables but also both secondary buses. Insulating material covered most of the bus, but the joints were exposed.

Will Suggs was an electrician for a maintenance-contracting firm in the Dallas area. PMC had issued Will's firm a contract to tape the open connections in each of the three 2,400-volt substations on the site. PMC had experienced a problem with small animals getting into the exposed joints.

On Wednesday, September 8, Will appeared at the vendor's gate at PMC to begin the work. After signing in and registering his truck, Will continued to the substation area and called for Craig on the public address system. Craig quickly responded, and Will indicated that he wanted to begin the contract at substation BA-5A1. Will left to gather his materials while Craig went about deenergizing bus B. Craig told Will that transformer B was down and locked out. Will asked which transformer was B and was told that B was the transformer on the right. Craig assumed that Will would know he meant on the right when looking at the front.

Will went around to the rear of the substation and opened the compartment door next to the transformer on his right. The bottom bus exited from that transformer. Will was certain that Craig knew that the work would be done from the rear of the transformer, so Craig must have meant on the right when looking from the work position. In fact, Will didn't even consider that Craig might have meant the transformer on the right, *looking from the front.*

Craig had some paperwork to do in the substation room. He went over to the desk and poured himself a cup of coffee, then began to make notes in the files.

Because he would be working on the bottom bus, Will saw no reason to cover any of the other exposed bus with insulating material. The load cable terminals were at least 24 inches away. He would not even be close to the other exposed bus, so he didn't cover anything.

Will's left arm was touching the cabinet as he reached to begin installing rubber tape on C phase. He made solid contact with the energized C phase conductor. As he worked at his desk, Craig thought that he heard a sound but returned to his paperwork.

Soon, Craig became concerned that he was hearing no sound from the back of the gear, and he looked around the corner. He saw Will's legs extending from the compartment. Craig called out to Will and got no response.

Instantly, Craig knew that his conversation with Will was not clear. There were no labels on the transformers. There was no mimic bus on the front of the switchgear. There was no bus isolation in the switchgear. Craig and Will had different plans. Neither Craig nor Will questioned the other's ability. Will had rejected minimizing exposure to the hazard. No voltage test was made.

Had either Craig or Will been thinking about his personal electrical safety principles, the incident could have been avoided. Instead, Will was dead on arrival at the local hospital. CPR had failed to resuscitate him. The death certificate suggested that the cause of death was electrocution. However, the real cause of this death could have been failure to apply personal principles.

Test Your Thinking

True	False	
❏	❏	1. Belief in a set of principles and use of those principles in daily work can help a person to expose a hazard or poor work practice.
❏	❏	2. As long as a workplace understands and follows OSHA and other published standards, individual workers do not need to plan for their own safety.
❏	❏	3. Because electrical hazards are very easy to see, a plan to review hazards before beginning electrical work is not necessary.
❏	❏	4. Once a good set of personal safety principles has been defined, they won't ever need to be changed.

Note

1. This account is based on an actual incident. The names, including the name of the facility, have all been changed to protect those involved. Any similarity to actual names or facilities is strictly coincidental.

Answers: 1. (true), 2. (false), 3. (false), 4. (false)

A Sample Electrical Safety Program

1.0 Scope

This procedure identifies the electrical safety program that should be in place covering all electrical work performed by _____ . This procedure provides overall program guidance and should be used in conjunction with all other procedures and practices employed on the site to help ensure that electrical work is accomplished safely.

2.0 General

2.1 Philosophy

All injuries are preventable. Sound safety practices are a condition of employment and of continued employment. Each person is responsible for avoiding exposure to a safety hazard and ensuring that all unsafe conditions are corrected. Each person is also responsible for identifying unsafe acts and for correcting such acts to the limit of his/her knowledge or ability.

2.2 Personal Responsibility

Each person is responsible for his or her own safety and for the safety of others. Each employee is expected to correct or report unsafe conditions of acts that are observed. Each person's attitude is reflected in his or her behavior. Each person is expected to know, understand, and use applicable safety procedures as tools to guide all tasks.

2.3 Controls

2.3.1 Shutting Down Energy Source

No work shall be accomplished where exposure to hazards associated with electrical energy exists until an attempt is first made to shut down the source of energy.

2.3.2 Parts Energized Until Proven Otherwise

Every electrical conductor or circuit part is considered energized until proven otherwise.

2.3.3 No Barehanded Contact

No barehanded contact shall be made with exposed energized conductors or circuit parts of more than 50 volts.

2.3.4 Deenergizing Is a Dangerous Task

Deenergizing an electrical conductor or circuit part and making it safe to work on is, in itself, considered a potentially hazardous task.

2.3.5 Provision and Application of Procedures

Each employer will provide procedures and each employee will apply them to accomplish each task.

2.3.6 Qualified Employees

Employees will be qualified for the task to which they are assigned.

2.3.7 Hazard/Risk Analysis

A hazard/risk analysis will be performed for each task involving any approach to energized conductors and circuit parts.

2.3.8 Overall Safety Environment

The overall safety environment will be considered when assigning personnel to tasks.

2.3.9 Safety Discussions
Each week will begin with a safety discussion.

2.3.10 Hazards and Procedures
Each job lineup will include a discussion of existing hazards and the procedures appropriate for the tasks involved in the job.

2.4 Standard of Performance

Practices and procedures in use by _____ employees are based upon criteria established in National Fire Protection Association Code NFPA 70E, *Standard for Electrical Safety Requirements for Employee Workplaces,* 2000 edition.

2.5 Training

Each site (and contractor) is expected to establish and execute a training program that includes information related to electrical hazards, how a person is exposed to electrical hazards, how a person protects himself or herself from exposure to injury, and how to use procedures to execute required work activities. The training program shall include annual knowledge review and procedural updating as standards and expectations change. A record of initial and review training shall be maintained.

2.6 Auditing

Each site (and contractor) shall establish an auditing process that reviews practices and conditions in the workplace. An audit shall be conducted at least weekly, and all unsafe conditions and unsafe acts shall be corrected expeditiously. Periodically, the audit shall target specific procedures for review.

2.7 Supervisory Responsibility

Each supervisor must set an example by demonstrating the proper attitude and behavior toward safety. The supervisor's conduct is reflected in the conduct of those he or she supervises. Each supervisor should empower the people under his or her direction to be proactive in continuously improving their own safety and the safety of others. Each supervisor shall ensure that the people under his or her direction have the necessary knowledge and skills to complete assigned tasks safely.

3.0 Policies

3.1 Standards Policy

National codes and standards that are to be implemented at _____ include NFPA 70, *National Electrical Code®*, 2000 edition), and NFPA 70B, *Recommended Practice for Electrical Equipment Maintenance*, 1998 edition.

Electrical safety related work practices implemented at _____ are guided by 29 *CFR* 1910, Subpart S, and 1910.147. In addition, electrical safety-related work practices shall meet all requirements contained in the latest edition of NFPA 70E.

3.2 As-Built Documentation Policy

Drawings used in planning electrical work must reflect the current condition of equipment and installations, Single-line diagrams, process and instrument (P&I) diagrams, schematics, and underground drawings must all be up-to-date so that proper planning can take place. In addition, up-to-date drawings help to identify potential hazards. Inaccurate drawings can compromise the safe execution of an electrical task, no matter how well planned the task might be. These drawings shall be maintained in an up-to-date condition. As-built changes shall be recorded, and file copies shall be changed appropriately.

Equipment shall be properly labeled and identified. As conditions change or revisions are made, equipment identification must be updated.

3.3 Abandoned Lines, Wires, or Cables Policy

Electrical lines, wires, and cables that are removed from service or not connected should be removed. If removal is not feasible, individual conductors must be taped and then tagged to indicate the location of the other end. Underground wiring that has been abandoned in place must be maintained in drawings for reference and so indicated on the drawing. Temporary wiring installed to provide power during construction must be removed when it is no longer required.

3.4 Excavation Policy

A thorough investigation must be conducted prior to beginning any excavation work. This investigation includes, but is not limited to, examining drawings and other documentation, requesting information from people in the area, and inspecting the area with devices that can detect underground obstacles.

■ Appendix B

Sample Procedures

Lock, Tag, Try, and Test

1.0 Purpose

This procedure establishes the minimum requirements for lockout of electrical energy sources. It is to be used to ensure that conductors and circuit parts are disconnected from sources of electrical energy, locked, and tested before work begins where employees could be exposed to dangerous conditions. Sources of stored energy, such as capacitors or springs, shall be relieved of their energy. A mechanism shall be engaged to prevent reaccumulation of energy.

2.0 Responsibility

All employees shall be instructed in the safety significance of the lockout procedure. All employees and all other persons whose work operations are, or might be, in the area shall be instructed in the purpose and use of this procedure. Each member of the line organization is responsible for ensuring that appropriate personnel receive instructions on their individual roles and responsibilities. All persons installing a lockout device shall sign and date the tag.

3.0 Preparation for lockout

To adequately prepare to implement a lockout, the qualified employee shall take the following steps:
1. Review current diagrammatic drawings (or other equally effective documents), tags, labels, and signs to identify and locate all disconnecting means to be locked.

2. Review disconnecting means to determine adequacy of the interrupting ability.
3. Determine if a visible break may be seen or if other precautions are necessary.
4. Review other work activity to identify where and how other personnel might be exposed to sources of electrical energy hazards.
5. Review other energy sources in the physical area to determine employee exposure to sources of other types of energy.
6. Define energy control methods for control of other hazardous energy sources in the area.
7. Obtain an adequately rated voltage detector to test each phase conductor or circuit part to ensure that they are deenergized.
8. Identify how to determine that the voltage detector is operating satisfactorily.
9. Where the possibility of induced voltage or stored electrical energy exists, ground the phase conductors or circuit parts before touching them.
10. Where it is possible that the deenergized conductor(s) could contact or be contacted by an exposed energized conductor or circuit part, install grounding devices.

4.0 Types of lockout

The following types of lockout are acceptable on this site:
- Individual employee control
- Simple lockout
- Complex lockout

4.1 Individual employee control

Individual employee control is defined as a method of energy control that is accomplished by operating a single isolating device. The single isolating device must be immediately accessible and continuously visible to the employee. The work must not extend beyond one work period.

Individual employee control is permitted without the placement of lockout devices on the isolating device.

The individual employee control procedure may be used when equipment is deenergized for activities such as minor maintenance, servicing, adjusting, cleaning, inspecting, and making corrections to operating conditions.

4.1.1 Implementing individual employee control

In addition to all the steps included in preparation for lockout (Section 3.0), the following steps shall be taken to implement individual employee control.

1. While wearing appropriate personal protective equipment, open the disconnect switch.
2. Open the door or remove the cover to the equipment.
3. Immediately inspect the interior of the open compartment for signs of impending failure.
4. Verify proper operation of the voltage-detecting device before testing for voltage.
5. Using the approved voltage-detecting device, test all conductors and circuit parts within the enclosure for absence of voltage.
6. Verify proper operation of the voltage-detecting device after testing for absence of voltage.

4.2 Simple lockout procedure

A simple lockout is defined as a lockout that involves the following:

- A single crew of employees
- A single source of energy
- Tasks that do not extend beyond a single work period.

After each step is covered in preparation for lockout (Section 3.0), each employee shall install his or her own lockout device on the isolating device.

A written plan is not required for simple lockout.

Each employee shall install his or her personal lockout device.

4.2.1 Implementing simple lockout

The employees shall be notified that a lockout system is going to be implemented and the reason therefore. The qualified person implementing the lockout shall know the disconnecting means location for all sources of electrical energy and the location of all sources of stored energy. The qualified person shall be knowledgeable of hazards associated with electrical energy.

In addition to all the steps included in preparation for lockout (Section 3.0), the qualified person shall take the following steps.

1. Deenergize the electric supply, if the electrical supply is energized, and relieve all stored energy.
2. Lockout all disconnecting means with lockout devices.
3. Attempt to operate the disconnecting means to determine that operation is not possible.
4. Use a voltage-detecting instrument. Inspect the instrument for visible damage. Do not proceed if any indication of damage to the instrument exists. Postpone the procedure until you procure an undamaged device.
5. Verify proper operation of the instrument on a known source.
6. Test for absence of voltage.
7. Verify proper operation of the instrument on a known source.
8. Where required, before touching them, install grounding equipment on the phase conductors or circuit parts to eliminate induced voltage or stored energy. Where it has been determined that contact with other exposed energized conductors or circuit parts is possible, apply ground devices rated for the available fault duty.

The equipment and/or electrical source are now locked out.

4.2.2 Procedure involving more than one person

For a simple lockout and where more than one person is involved in the job/task, each person shall install his or her own personal lockout device.

4.2.3 Procedure involving more than one shift

When the lockout extends for more than one day, the lockout shall be verified to be still in place at the beginning of the next day. As long as the same crew is back the following day to continue the work, the job is still considered a simple lockout. Where the lockout is continued on successive shifts, the lockout is considered to be a complex lockout because different workers are involved.

4.3 Complex lockout

4.3.1 Where complex lockout is required

A complex lockout plan is required where one or more of the following exist:

- Multiple energy sources (more than one)
- Multiple crews
- Multiple crafts
- Multiple locations
- Multiple employers
- Unique disconnecting means
- Complex or particular switching sequences
- Continuation of the lockout for more than one shift (thus involving workers who did not participate in the original lockout)

4.3.2 The person-in-charge

A person-in-charge shall be involved with a complex lockout procedure. The person-in-charge shall do the following:

1. Develop a written plan of execution.
2. Communicate the plan of execution to all persons engaged in the job/task.

3. Be held responsible for the safe execution of the complex lockout plan.
4. Ensure that each person understands the hazards to which they are exposed and the safety-related work practices that are to be used.

The person-in-charge can install locks or direct their installation on behalf of other employees.

The person-in-charge can remove locks or direct their removal on behalf of other employees only after all personnel are accounted for and ensured to be clear of potential hazards.

Where the complex lockout is continued on successive shifts, the person-in-charge shall identify the method for transfer of the lockout and of communication with all employees.

At this location, _____ shall be the person-in-charge.

4.3.3 The complex lockout plan

The complex lockout plan shall do the following:
1. Address all the concerns of employees who might be exposed and ensure that they understand how all electrical energy is controlled.
2. Include all steps that are necessary to prepare for the lockout.
3. Identify all steps that are necessary to implement the required control of exposure to electrical hazards.
4. Identify all steps that are necessary to restore the energy to normal condition
5. Identify the method to account for all persons who might be exposed to electrical hazards in the course of the lockout.

4.3.4 Method to be used

The lockout method to be used should be selected from the following:

- Each individual shall install his or her own personal lockout device.
- The person-in-charge shall lock his or her key in a "lock box" and maintain a sign-in/-out log for all personnel entering the area.
- The person-in-charge shall select an equally effective method.

5.0 Restoring the equipment and/or electrical supply to normal condition

After the job/task is complete, the qualified employee or person-in-charge shall do the following:

1. Visually verify that the job/task is complete.
2. Remove all tools, equipment, and unused materials and perform appropriate housekeeping.
3. Notify all personnel involved with the job/task that the lockout is complete, that the electrical supply is being restored, and that they must remain clear of the equipment and electrical supply.
4. Perform any quality control tests/checks on the repaired/replaced equipment and/or electrical supply.
5. Order removal of the lockout devices by the person who installed them.
6. Return the disconnecting means to their normal condition.

6.0 Discipline

Knowingly violating this procedure will result in _____ (State disciplinary action to be taken.)

Knowingly operating a disconnecting means with an installed lockout device will result in _____ (State disciplinary action to be taken.)

7.0 Testing for voltage

In every instance before contact, each conductor or circuit part shall be tested for voltage with an approved voltmeter. The voltmeter shall be tested for proper operation on a known energy source both before and after the test.

8.0 Equipment

Locks shall be _____ (State type and model of selected locks.)

Tags shall be _____ (State type and model to be used.)

Voltage detecting device(s) to be used shall be _____ (State type and model.)

9.0 Review

This procedure was last reviewed on _____ (state date of last review.) This procedure will be reviewed again on _____ (State date of next review.)

10.0 Lockout training

Training that includes the following shall be provided for all qualified employees:
• Recognition of lockout devices
• Installation of lockout devices
• Duty of employer in writing procedures
• Duty of employee in executing procedures
• Duty of person-in-charge
• Authorized and unauthorized removal of locks and tags
• Enforcement of execution of lockout procedure
• Individual employee control procedure
• Simple lockout procedure
• Complex lockout procedure
• Use of single-line and diagrammatic drawings to identify sources of energy

- Use of tags and warning signs
- Release of stored energy
- Personnel accounting methods
- Grounding needs/requirements
- Safe use of voltage-detecting devices

Creating an Electrically Safe Work Condition

1.0 Purpose

The purpose of this practice is to establish minimum actions necessary to establish an electrically safe working condition.[1] This practice must be used in conjunction with a lock, tag, try, and test procedure for electrical energy sources. It should be used to ensure that personal exposure to electrical hazards is avoided where possible and minimized where avoidance is not possible.

2.0 Responsibility

In every possible instance, each employee is responsible for implementing this practice prior to beginning any work where possible exposure exists to hazards associated with electricity.

3.0 General

NOTE: Opening a disconnecting means and inspecting within an enclosure is a potentially hazardous task. A hazard/risk analysis should be performed to identify potential exposure.

Any electrical circuit or conductor is considered energized until the source of energy is removed, at which time the circuit or conductor shall be considered deenergized. Any electrical circuit conductor and circuit part shall not be considered to be in an electrically safe work condition until all of the following steps have been completed.

1. Identify all possible sources of electrical energy.
2. Open the disconnecting devices of each source.
3. Where possible, visually verify all blades are open or breakers have been withdrawn to the fully disconnected position.
4. Apply lockout/tagout devices.
   NOTE: This step (number 4) is not required for individual qualified employee control.
5. Use a voltage detector to verify that all conductors and circuit parts are deenergized.
6. Apply discharge means or safety grounds, as required.

[1] NFPA 70E, Part II.

All work on or near electrical equipment not placed in an electrically safe condition shall use safe work practices appropriate for the voltage and energy level.

4.0 Stored energy

Hazardous energy may exist as stored energy in many instances. Stored energy must be discharged or otherwise relieved prior to blocking and installing lockout devices.

Capacitors must be discharged, shorted, and grounded in addition to locking out the source of energy. Springs shall be released or a physical restraint applied when necessary to immobilize mechanical, pneumatic, or hydraulic equipment. Other sources of stored energy shall be blocked or relieved.

5.0 Voltage-detecting instruments

All voltage-detecting devices shall be rated for the voltage and be suitable for the environment in which it will be used. To test for the presence or absence of voltage, a single-function voltage-detecting device should be used. Proper operation of the voltage tester must be verified on a known source, both *before* and *after* the tests.

6.0 Preparation

These steps should be followed before an electrically safe work condition can be created.

1. Ensure that record drawings are up-to-date and available to employees.
2. Ensure that a process exists to bypass circuit interlocks.
3. Make a list of energy-isolating devices to be locked/tagged.
4. Identify energy-isolating devices requiring additional precautions.
5. Identify the appropriate voltage detectors required for testing for the absence of voltage at each location to be tested and the method that will be used to determine if the voltage detector is working properly.

Testing for Absence of Voltage —1,000 Volts and Below

1.0 Basic safety issues

Three basic safety issues are associated with the task of testing for voltage in instances where the maximum voltage level is 1,000 volts and below. The first issue involves selecting and using the right meter for the job at hand. The second issue is protecting the person from potential exposure to an energized source, and the third issue is the work process of executing the test.

1.1 Selecting a voltage-testing device

1.1.1 Establishing an electrically safe work condition

Voltage testers should be selected based upon the intended use. Several types of voltage testers are manufactured for specific uses, and each device has limitations. When used to test for the absence or presence of voltage as a part of establishing an electrically safe work condition, voltage testers should have the following characteristics where direct contact can be made:

- Retractable, insulated-tip test probes
- Self-contained fault protection or limitation devices, such as internal current-limiting fuses or probe current-limiting resistors
- Voltage/current path from the probes that is not routed through the mode switch

In addition, voltage testers should conform to national consensus standards, such as UL 1244, MIL-T-28800C.

1.1.2 Testing only for absence or presence of voltage

Along with the above requirements, voltage testers that are used only to test for the absence or presence of voltage should have the following characteristics:

- Single-function, voltage-only test devices or automatic mode test devices that check for voltage before switching to other modes (i.e., resistance, continuity)
- Test leads that cannot be improperly connected (i.e., only two jacks are present or leads are permanently connected)

  NOTE: High-impedance voltage testers are subject to "phantom" readings from induced voltage. Verification of the absence of voltage may be required with a low-impedance voltage tester, such as a solenoid-type voltage tester. Solenoid testers may have an adverse effect on digital control systems (DCS), programmable logic controllers (PLC), or similar equipment.

  NOTE: Solenoid-type voltage testers typically are assigned a "duty cycle" by the manufacturer. In most instances, this duty cycle is 15 seconds. The "duty cycle" rating must not be exceeded.

1.1.3 Selecting voltage-testing devices that minimize injury possibility

On occasion, voltage-testing devices can be the source of an incident or injury, as in the following instances:

- Leads can fall out of their plugs and initiate a phase-to-phase short circuit.
- Internal components can fail, resulting in a phase-to-phase short circuit.
- Probes can slip while a reading is being observed.
- Leads can be inserted into incorrect plugs, resulting in failure.
- The device indication can be confusing, resulting in incorrect observations.
- Hands can slip off the probe.

The selected voltage-testing device must minimize all of these possibilities.

1.2 Protecting the person

Prior to opening doors or removing covers for access to electrical conductors, a person should conduct a hazard analysis. The hazard analysis should be as formal and detailed as warranted by the task to be performed. Any personal protective equipment (PPE) necessary to avoid injury should be in place and worn before any existing enclosure is abridged, i.e., removing any cover or opening any door. The hazard analysis must consider both shock and arc flash.

NOTE: Many arc-flash incidents occur at the moment a door is opened or a cover removed. The person performing the test should be aware of this fact and exhibit an appropriate mind set. The mind set should consider that all electrical conductors and contact points within the enclosure are energized.

In determining appropriate PPE, the hazard analysis must consider the arc-flash boundary as well as the shock approach boundaries, paying particular attention to the prohibited and restricted boundaries. Where the task involves measuring a voltage, the probes, of course, cross the prohibited boundary. Therefore, the person must be protected from unintended contact with conductive parts.

Voltage-testing devices that meet the above criteria include a preventive method to minimize the possibility of a person's hand or fingers slipping down the probes. Therefore, electrical insulation is not necessarily required. However, if hands (or other body parts) are inside the enclosure while the person is executing the task, some exposure to shock exists through unintentional contact with energized or potentially energized parts. Voltage-rated gloves should be worn. They do not hinder the task and can avoid unintentional contact with electrical conductors or contacts.

In every instance where an electrical circuit is present, an arc-flash boundary exists. Depending upon the arc-flash boundary (as calculated by the Lee method contained in NFPA 70 E,1995 edition), flash-protective equipment should be worn. Any body part that is within the arc-flash boundary must be protected from arc flash. If the arc-flash boundary is

2 inches or less, leather gloves and ordinary safety glasses for the eyes provide sufficient protection. As the arc-flash boundary extends beyond 2 inches, flame-resistant clothing and face protection should be worn.

Leather gloves that are one component of voltage-rated gloves provide arc-flash protection for hands. Therefore, appropriate voltage-rated gloves should be worn. *Voltage-rated gloves selected in accordance with ASTM D 120 provide protection from both shock and arc flash, in most instances.*

1.3 Executing the task

The person testing for voltage should be trained to understand how the meter works and what each possible meter indication means. After the person selects the appropriate volt meter, reacts to the hazard analysis, and understands how to interpret any meter indication, he or she should execute the following sequence of steps:

1. Open the disconnecting means.
2. Open door or remove cover(s).
3. Inspect the compartment interior for missing barriers, signs of arcing or burning, and any extraneous parts or components.
4. Inspect the voltmeter and probes for signs of mistreatment; verify that the probe covers move freely.
5. Insert one probe into the holder on the meter; place the meter in a stable position or ask a second person to hold the meter, if necessary, to see the indication. (Any second person must wear the same PPE as the first person.)
6. Verify that the voltmeter functions satisfactorily on a known energized voltage source.
   NOTE: If the meter is auto ranging, a nearby 110-volt receptacle is satisfactory. If not auto ranging, the known source must be within the same voltage range.
7. Place the probe that is in the meter holder into good physical contact with a grounded point within the compartment.
8. Place the second probe into good physical contact with

the opened side of the disconnecting means and before (ahead of) any fuses or any other circuit element.

NOTE: Normally, in the case of a disconnect switch, the movable side of the knife blades are available to contact with the probe. In the case of a circuit breaker, the load conductor termination should be contacted.

9. Read and interpret the meter indication.
10. Repeat steps 7 and 8 for phases B and C.
11. Place the probe that is in the meter holder into good physical contact with phase A on the opened side of the disconnecting means and before (ahead of) any fuses or other circuit element.

    NOTE: Normally, in the case of a disconnect switch, the movable side of the knife blades is available to contact with the probe. In the case of a circuit breaker, the load conductor termination should be contacted.
12. Place the probe in the meter holder into good physical contact with phase B in the same relative physical location.
13. Repeat steps 11 and 12, except contact phases B and C.
14. Repeat steps 11 and 12, except contact phases A and C.

    NOTE: Tests for absence of voltage should be conducted at each point within the enclosure. If the compartment contains fuses, a voltage test should be conducted at both the line and load sides of each fuse, both between phases and between each phase conductor and ground. Each test should be taken at the fuse clip instead of at the fuse.
15. Measure voltage between each point within the enclosure where contact is expected.
16. Verify that the voltmeter functions satisfactorily on a known energized voltage source.

    NOTE: If the meter is auto ranging, a nearby 110-volt receptacle is satisfactory. If the meter is not auto ranging, the known energized source must be within the same voltage range.

When a voltage test is performed, the person should perform the work practice as if the energy source is present (source is energized). Even if the disconnecting means has been opened, until the absence of voltage has been satisfactorily verified, a safe work condition does not exist. The person performing the test should be protected from any accidental release of energy until the absence of voltage has been satisfactorily verified.

2.0 Troubleshooting tips

2.1 Resetting overload relays

When an overload relay operates, fuses blow or circuit breakers trip for unknown reasons, they may be reset or replaced one time. If they trip the circuit a second time, the absence of a faulted condition must be verified by circuit examination prior to replacing or resetting the circuit element, according to OSHA [29 *CFR* 1910.304 (b)(2)].

2.2 Checking fuses for continuity

When it is suspected that a fuse has opened, both indicating and nonindicating fuses should be removed from the circuit and checked for continuity. In some knife-blade fuse constructions, both the fuse barrel and the ferrule (endbell) are insulated. Care must be taken to make certain that any measurements are taken from the uninsulated portion of the device such as the "knife" instead of the ferrule.

2.3 Troubleshooting on deenergized equipment

To minimize exposure to electrical hazards, troubleshooting should be performed on deenergized equipment, where possible. Resistance measurements are as reliable as voltage measurements.

Bibliography

ANSI Z 87.1, *Practice for Occupational and Educational Eye and Face Protection.* New York: American National Standards Institute, 1989.

ANSI Z 89.1, *Requirements for Protective Headwear for Industrial Workers.* New York: American National Standards Institute, 1997.

ASTM F 496, *Standard Specification for In-Service Care of Insulating Gloves and Sleeves.* Philadelphia: American Society of Testing and Materials, 1997.

ASTM F 696, *Standard Specification for Leather Protectors for Rubber Insulating Gloves and Mittens.* Philadelphia: American Society of Testing and Materials, 1997.

ASTM F 855, *Standard Specification for Temporary Grounding Systems to Be Used on De-energized Electric Power Lines and Equipment.* Philadelphia: American Society of Testing and Materials, 1997.

ASTM F 1117, *Standard Specification for Dielectric Overshoe Footwear.* Philadelphia: American Society of Testing and Materials, 1993.

ASTM F 1503, *Standard Practice for Machine/Process Potential Study Procedure.* Philadelphia: American Society of Testing and Materials, 1998.

ASTM F 1505, *Standard Specification for Insulated and Insulating Hand Tools.* Philadelphia: American Society of Testing and Materials, 1994.

ASTM F 1506, *Standard Specification for Protective Wearing Apparel for Use by Electrical Workers When Exposed to Momentary Electric Arc and Related Thermal Hazards.* Philadelphia: American Society of Testing and Materials, 1998.

Capelli-Schellpfeffer, Mary. Personal correspondence, Sept. 29, 1999.

Capelli-Schellpfeffer, Mary. "What Can Management Do?" in Electrical Injury: A Multidisciplinary Approach to Therapy, Prevention, and Rehabilitation. *Annals of the New York Academy of Sciences. Annals of the New York Academy of Sciences,* edited by Raphael C. Lee, Mary Capelli-Schellpfeffer, and Kathleen M. Kelley, Vol. 720. New York: The New York Academy of Sciences, 1994.

Capelli-Schellpfeffer, M, M. Toney, R. C. Lee, and R. D. Astumian. "Advances in the Evaluation and Treatment of Electrical and Thermal

Injury Emergencies." Paper presented at the Forty-First Annual Conference of the IAS/IEEE Petroleum and Chemical Industry Committee, Vancouver, British Columbia, September 12–14, 1994.

Capelli-Schellpfeffer, Mary, Raphael C. Lee, Mehmet Toner, and Kenneth R. Diller. "Correlation between Electrical Accident Parameters and Sustained Injury." Paper presented at the Forty-Third Conference of the IAS/IEEE Petroleum and Chemical Industry Committee, Philadelphia, PA, September 23–25, 1996.

Capelli-Schellpfeffer, Mary, Christine Kalina, Michael K. Toney, John H. Mitchell, and Raphael C. Lee. "Partnerships for Electrical Safety." Paper presented at the Forty-Sixth Annual Conference of the IAS/ IEEE Petroleum and Chemical Industry Committee, San Diego, CA, September 13–15, 1999.

Crawford, Kenneth S., David G. Clark, and Richard L. Doughty. "Motor Terminal Box Explosions Due to Faults." *IEEE Transactions*, Vol. 29, No. 1, Jan./Feb. 1993.

Doughty, R. L., R. A. Epperly, and R. A. Jones. "Maintaining Safe Work Practices in a Competitive Environment," *IEEE Transactions*, Vol. 27, No. 5, Sept./Oct. 1991.

Doughty, R. L., T. E. Neal, and H. L Floyd. "Predicting Incident Energy to Better Manage the Electrical Arc Hazard on 600 V Power Distribution Systems." Paper presented at the Forty-Fifth Annual Conference of the IAS/IEEE Petroleum and Chemical Industry Committee, San Diego, CA, September 28–30, 1998.

Fatal Injuries to Workers in the United States, 1980–1989: A Decade of Surveillance. Cincinnati, OH: U.S. Department of Health and Human Services, Centers for Disease Control and Prevention, National Institute for Occupational Safety and Health. DHHS (NIOSH), Publication 93–108, 1993.

"Improving the Diagnosis and Treatment of Electrical Burns." *EPRI Journal*, Vol. 21, No. 3, May/June, 1996, pp. 4–5.

Jones, Jane G. "Report on Electrical Shock and Burn Treatment Captivates Audience at PCIC Safety Session." *IEEE Industry Applications Magazine*, Vol. 1, No. 2, March/April 1995.

Jones, Ray A., et al. "Staged Tests Increase Awareness of Arc-Flash Hazards in Electrical Equipment." Paper presented at the Forty-Fourth Conference of the IAS/IEEE Petroleum and Chemical Industry Committee, Banff, Alberta, Canada, September 15–17, 1997.

Lee, Raphael C. "Injury by Electrical Forces: Pathophysiology, Manifestations, and Therapy." *Current Problems in Surgery*, Vol. 34, No. 9, Sept. 1997, pp. 677–765.

Lee, Raphael C., Mary Capelli-Schellpfeffer, and Kathleen M. Kelley, eds. *Electrical Injury: A Multidisciplinary Approach to Therapy, Prevention, and Rehabilitation. Annals of the New York Academy of Sciences*, Vol. 720. New York: The New York Academy of Sciences, 1994.

Lee, Raphael C., Adam Myerov, and Christopher P. Maloney. "Promising Therapy for Cell Membrane Damage." *Electrical Injury: A Multidisciplinary Approach to Therapy, Prevention, and Rehabilitation. Annals of the New York Academy of Sciences. Annals of the New York Academy of Sciences*, Vol. 720. New York: The New York Academy of Science, 1994.

Lee, R. H. "The Other Electrical Hazard: Electric Arc Blast Burns." *IEEE Transactions*, Vol. IA–18, No. 3, May/June 1982.

Maslow, A. H. *Motivation and Personality* (2nd ed.). New York: Harper & Row, 1970.

Meadowcroft, W. H. *The ABC of Electricity*. New York: Home Book Company, 1988.

Neal, Thomas E., Allen H. Bingham, and Richard L. Doughty. "Protective Clothing Guidelines for Electric Arc Exposure." Paper presented at the Forty-Third Annual Conference of the IAS/IEEE Petroleum and Chemical Industry Committee, Philadelphia, PA, Sept. 23–25, 1996.

NFPA 70, *National Electrical Code®*. Quincy, MA: National Fire Protection Association, 1999.

NFPA 70B: *Recommended Practice for Electrical Equipment Maintenance*. Quincy, MA: National Fire Protection Association, 1998.

NFPA 70E: *Standard for Electrical Safety Requirements for Employee Workplaces*. Quincy, MA: National Fire Protection Association, 2000.

The OSH Act: Public Law 91-596. Ninety-First Congress of the United States of America, S2193, Dec. 29, 1970.

OSHA Inspection Instruction STD 1.73. Washington, DC: Occupational Safety and Health Administration, U.S. Department of Labor.

OSHA Regulations 29 *CFR* 1900 to 1910. Washington, DC: Occupational Safety and Health Administration, U.S. Department of Labor.

OSHA Regulations 29 *CFR* 1926. Washington, DC: Occupational Safety and Health Administration, U.S. Department of Labor.

Parker, Steve. *Thomas Edison and Electricity*. New York: HarperCollins, 1992.

Possible Health Effects of Exposure to Residential Electric and Magnetic Fields. Washington, DC, National Academy Press, 1996.

San Francisco State University Home Page http://www.sfsu.edu/~markd/acpower.htm, *April 8, 1999*.

Sargent, Jeffrey S. and Noel Williams. *NFPA Electrical Inspection Manual with Checklists*. Quincy, MA: National Fire Protection Association, 1999.

UL 943, *Ground-Fault Circuit Interrupters*. Northbrook, IL: Underwriters Laboratories, 1993.

U.S. Department of Energy, *Electrical Safety Guidelines*, Appendix A, Sept. 1993.

Worker Deaths by Electrocution: A Summary of Surveillance Findings and Investigative Case Reports. Cincinnati, OH: U.S. Department of Health and Human Services, Centers for Disease Control and Prevention, National Institute for Occupational Safety and Health. DHHS (NIOSH), Publication 98–131, 1998.

Index

Abandoned lines, wires or cables policy, sample, 266

Accidents. *See* Incidents

Accountability, safety, 3–4

AHJ (authority having jurisdiction), 102, 107

American National Standards Institute (ANSI), 99

American Society for Testing Materials (ASTM) standards
  insulated hand tools, 184
  personal protective equipment, 30, 76, 77*t*, 78*t*
  temporary grounding, 24

Apprentice programs, 66

Approach boundaries, 188–190, 189*f*

Arc blast
  description of, 32
  recognizing, 32–33
  risk management, 56

Arc flash, 61
  addressing, 33–35
  burns from, 22–23, 199–200
  degree of exposure, reducing, 31–32
  description of, 22
  example of, 225–226
  eye/face protection for, 80–81
  head protection for, 79
  in metal enclosure, 33
  personal protective equipment for, 81–84, 82*f*, 84*t*
  plasma, 62, 63*f*
  pressure
    measurements of, 35*f*
    results from, 62, 63*f*
  risk management, 56

Arc-flash boundary, 54

Arc-flash fault, 29

Arc-flash testing, 30

Arc-resistant switch gear, 220

Area classification
  for risk management, 54–55

As-built documentation policy, sample, 266

ASTM standards. *See* American Society for Testing Materials (ASTM) standards

Audit/auditing
  behavior, 249–250
  conducting, 246–247
  documentation of, 248–249
  effective, result of, 250
  for electrical safety program, 151–152
  examples of, 250–251
  follow-up, 247–248
  interm, conduction of, 248
  objective of, 245
  preparation of, 246
  purpose, 147, 245–250
  records, 147–148
  training, 248–249

Authority having jurisdiction (AHJ), 102, 107

Available bolted-fault current (MVA$_{bf}$), 27

Awareness, of hazards, 179, 218

Barricades, 181–183

Barriers, 61. *See also* Personal protective equipment
  boundaries concept, 188–190, 188*f*, 189*t*

Behavior
  auditing, 249–250
  condition/arrangement of physical area and, 159–160
  example of, 163–165
  implications, 157–158
  motivation and, 160–162, 160*f*
  tolerance for risk taking, 159
  unsafe acts and, 162–163, 163*f*
  in workplace environment, 158–159

Blind reaching, 178–179